Nigel Hoffmann PhD has for
school teacher, in Australian
schools. He is the author of *Go*
The Artistic Stages (Adonis Press) and is a director of the
Education for Social Renewal Foundation.

THE UNIVERSITY
AT THE THRESHOLD

Orientation through Goethean Science

Nigel Hoffmann

RUDOLF STEINER PRESS

Rudolf Steiner Press
Hillside House, The Square
Forest Row, East Sussex RH18 5ES

www.rudolfsteinerpress.com

Published by Rudolf Steiner Press 2020

© Nigel Hoffmann 2020

A catalogue record for this book is available from the British Library

ISBN 978 1 85584 583 1

Cover by Andrew Morgan Design
Typeset by Symbiosys Technologies, Visakhapatnam, India
Printed and bound by 4Edge Ltd., Essex

Contents

Acknowledgements

First and foremost I want to acknowledge my wife Luisa's contribution which took the form of many conversations, over many years, about the future of the university. Thanks go to Dr Dalia Nassar and Dr Luke Fischer for suggestions on philosophical matters and to Dr Malte Ebach for advice on aspects of zoology. Thanks also to John Barnes for his very many valuable suggestions. This publication was made possible through financial support from Catherine Pilko, Alf Finch, Peter Want and the Anthroposophical Society in Australia.

Introduction

Slowly and seemingly inexorably, we are destroying life on Earth. We live in the proximity of a great danger, a danger which has only really dawned on humanity broadly in the last twenty years. Our situation is disquieting but it has a redemptive aspect, for it heightens the care and love people feel for nature and human life. Care springs from the whole human being—the thinking, heart and will—and is implicit in the scientific method of conscious inner participation in nature which derives from the work of Johann von Goethe, the German poet and scientist. The Goethean approach is not an alternative to modern science but rather—in the sense which is articulated in this book—a complement to and further development of that science. What is being advanced on these pages is the view that our situation today calls for a way of studying and understanding the world, of teaching and researching, which is an expression of our profoundest care. Far from being something incidental or peripheral to university education, the Goethean approach can provide an orientation to all studies at the tertiary level.

With respect to the concerns and cares of the younger generation, the question of the university is central. So to begin with it is necessary to build a thought-picture of the place and role of the university in the living body of society. The university appears as a 'little city,' focussing and intensifying those strivings towards understanding which are otherwise spread out, as it were, in the aspirations of individuals throughout the social fabric. The university expresses the quintessential human responsibility of marrying spirit and matter; that is to say, through reaching into the realm of spirit or mind, through inspired, creative thinking, the university serves to bring about new understandings in relation to new needs—this is the research dimension. And through decisive, imaginative deeds of the spoken and written word, to manifest and make effectual what has been understood—that is the teaching aspect.

Seen in this way it becomes possible to understand why the university can and should represent the ideal of human freedom in our time. Full responsibility in researching and teaching can only come about in a situation of freedom from political and economic imposition or obligation; this is why we expect so much of our universities as central organs of the secular state. It is vital that our universities are responding in freedom and with understanding to the deepest needs of our age but the reality is that discerning what these needs are is not an easy task and many obstacles may hinder the researcher.

Today we assume that the path of scientific knowledge leads simply from few facts to many facts, from elementary to more sophisticated theories, the stated aim always being to 'expand human knowledge'. The tool by which this path is forged is the analytic research process, powered by mechanistic logic, and applied to the inorganic and organic worlds alike. Scarcely is pause given for the consideration that the analytic process may be insufficient to understand living form, that what is alive must be approached in a way altogether different to that which is lifeless. Only out of concern to understand this difference will the true science of living form enter the mainstream of thought and the university curriculum. The philosopher Thomas Kuhn spoke of paradigm shifts in science, meaning developments which are not linear and continuous.[1] However this does not adequately express the transition from inorganic to organic thinking. The way into the sphere of living form requires more than an intellectual shift; it calls for the transformation of the actual *quality* and gesture of thinking. The word 'threshold' speaks more truthfully and meaningfully of this transition.

A threshold is a juncture of the path unlike any other, although many events may have anticipated it. What occurs there is the realization that what is beyond is utterly different from what has already been passed through. The threshold demands a change of heart, a turning around in the seat of consciousness. To stand at the threshold means that there must at least be the will and potential to pass beyond. It is an experience which has always been grasped by human beings in all times, in one way or another. In traditional lore thresholds are symbolically guarded, as with the rows of sphinxes and colossal pharaonic figures guarding the threshold formed by the stern pylons of the ancient Egyptian temples. Across the ancient world there was a vast pantheon of liminal deities. What these guardians symbolize is no light matter for human intelligence; a great demand is made and a great question hovers over the aspirants and tests them to the core. Who can say that we do not still have such stern demands on the path of scientific knowledge today? In truth, we do—although this is something which is scarcely recognized. In practice the scientific researcher faces great hindrances on the path which have to be overcome if progress is to be made; four of these, as identified by Goethe, are ease, self satisfaction, rigidity and prejudice.[2] At the threshold to the sphere of living form the disposition of the aspirant and the path itself are cast in a wholly new light.

In the early modern period the approach to this threshold was marked by the cultural polarization and tension between the stream of Enlightenment thinking, connected to the Industrial Revolution and exemplified by the work of Isaac Newton, and the ideas and insights of the philosophers, artists and natural scientists who are conventionally called 'the Romantics' (although this name has unfortunate connotations of sentimentalism.) With the latter first arose what could be called a culture of concern. The contemporary environmental movement grew out of the work of these 'Romantics,' who include the English poets Coleridge and Wordsworth, and the American Transcendentalists Emerson and Thoreau. These thinkers were concerned in one way or another about the progress and application of knowledge; they sensed with the 'dark Satanic Mills' of the Industrial Revolution that the living being of nature and the human being were physically and spiritually imperilled.[3]

In our time a far more developed culture of concern has brought humanity closer to the threshold of the sphere of living form, impelled in no small degree by the chaos visited upon us by the world wars of the twentieth century and more recently by the pandemic and 'great lockdown' of 2020. The global nature of the contemporary environmental and political-economic crises even casts doubt upon the future of the planet. Concern is felt by individuals in all nations and it is certainly being responded to by universities and governments worldwide. However, to a large extent this response only takes the form of technological and political countermeasures. There is also great concern that what we do technologically and politically is not enough and cannot ever be enough to overcome our problems, that a deeper transformation is required, that we need something more than what any new political strategy or technological breakthrough is able to achieve.

But there is something more than concern which lives within the worldwide environmental movement—it is *care*. Care is a quality of human feeling and will which arises out of concern the way a flower arises out of the vegetative plant. Care is a blossoming of concern. Concern appraises and responds to an objective situation; care embraces that situation nurturingly and protectively. Care radiates the light and warmth of conscience on everything from genetic engineering to nuclear technology to economic inequality and political abuse—on the whole life-field of the Earth in which every individual living being inheres. This care is no soft-minded wish to halt human progress; it is tested by the sternest requirement for clear-sightedness and commitment. Only care can engender within the human mind and will the power to hear and meet the demand for genuineness and self-knowledge at the threshold. Concern allows us to approach the threshold—but it is care which makes possible the crossing because through care, as in an act of love, we sacrifice our self-preoccupation for the sake of another being.

Goethe's way of science is the fruit of an eminent creative individual who stood at the threshold between the inorganic and the organic and through great care made the crossing, at least to a degree. Goethe, like the other 'Romantics', was concerned that mechanistic scientific ways of thinking would encroach upon the living being of nature and extirpate what he most revered. Concern which became care is what led Goethe to ask these kinds of questions of the work of the scientist and artist: What is an organism and how do we come to understand it *on its own terms*? What is the science of living form in contradistinction to the science of mechanics? What are the unique laws and forces which belong to living things? Goethe was significant because, by virtue of his particular capacities in which the scientific and artistic were naturally united, he was able to shape a new form of science capable of understanding life through conscious, inward participation in living form. It is for this reason that we speak of Goethe's artistic way of science. Every cultural movement has its forerunner, its initiator, and the work of Goethe is now two centuries in the past. Since his time the modern world has come to provide the conditions in which care could grow to become the heart and will impulse of the large part of humanity who comprehend our grave environmental and political-economic circumstances. This is the medium, we could say, in which Goethe's artistic way of science will now be able to be more broadly understood.

The philosopher Martin Heidegger has illuminated the following line of verse by the German lyric poet Hölderlin: 'But where there is danger, there the power to save also grows'.[4] Heidegger comments that the danger and the saving power are one and the same thing in the sense Hölderlin intends. When we truly understand the danger, the saving power will arise in us in response. Heidegger asks: 'What does "to save" mean? It means . . . to take under one's care, to keep safe.' In the first decades of the twenty-first century the danger is clear, largely unmitigated and growing. Yet the saving power is close at hand, but we hardly realize that it has a great deal to do with our scientific approach to the world. Concern for the world today provides the impetus to ask of ourselves a profound question which springs from care and from the will to keep safe the beings for whom we are responsible: *how can our way of knowing, the very style of our thinking which informs our research and our teaching, come to express care, to reveal itself to be a deed and duty of care?* This question is the threshold before which the contemporary university stands.

Part I
The Metamorphosis of The University

'. . . a true development of education must tend to nothing less than a superseding of [the] "Doctor" principle'.

Rudolf Steiner[5]

'For the inorganic, we have mathematics, geometry; for the organic, the living, we possess, to begin with, nothing that is inwardly so formed as, for example, a triangle, a circle, an ellipse. We attain to this [investigation of the organic aspect of nature] by means of living thought, not with the ordinary mathematics of numbers and figures, but with a higher mathesis, a mathesis that is qualitative, that works formatively, that reaches upward into the artistic . . . '

Rudolf Steiner[6]

'Love does not dominate, it cultivates. And that is more.'

Goethe[7]

'Goethe made the interesting statement that one should never form judgements or hypotheses about external phenomena because the phenomena themselves are the theories, they themselves give expression to their ideas if one has made oneself mature enough to allow them to work upon one in the right way. It is not a matter of making an effort to squeeze out of one's soul what one considers to be correct; one must be prepared for the judgement to leap towards one out of the facts themselves. Our relationship with thinking must not be to make thinking the judge of things but to allow it to be an instrument through which the things speak to us.'

Rudolf Steiner[8]

Chapter 1
Towards The Threshold of Living Form

Care

These pages explore the metamorphosis of the university beyond the centuries-old doctoral form, a development inspired by the educational ideal of the 'whole human being' or 'universal human'. Goethean science, a science of the whole human being, has a foundational role in the realization of the new university; for, as set forth here, this science forms the basis of an orientation course of studies at the heart of this university. An individual's entire educational development in relation to the different specializations can then be shaped out of an experience of the fundamentals of this science.

The 'doctor' principle of education, insofar as it is expressed in modern science, utters its aims in familiar ways: it speaks of illumination, of the search for truth and the virtue of 'objective, reliable knowledge'. The issue is not about turning away from the demands of truth, only about enhancing our understanding and experience of truth so that we may enter with insight and responsibility into the sphere of living form. Scientific objectivity means the striving for hard factuality and attaining the quality of clear seeing and unbiased rationality; this is a vital stage in knowing. In the method of Goethean science we can work with each quantum of analytical or factual knowledge in a very particular way if—and this 'if' is the key point—if in thinking we are able to negotiate the threshold to the realm of living form. Here the power of cognitive imagination can take hold of this knowledge as a kind of revitalizing force. The living being of nature, which in a sense 'dies' in abstract theoretical knowing, is brought to a new life through the cognitive imagination.

For students of university age, the educational ideal of the whole human being touches the heart and conscience in a particular way—it speaks directly to their sense of care. It is care for the world which brings us to the threshold of the sphere of living form and makes us worthy of the crossing. Care wishes to enhance the doctoral ideal of teaching and learning—centuries old and so fruitful in human development—with an understanding that is deeper and engages the whole human being. This book involves a critique of the doctoral form of education but this doesn't mean that the achievements of doctoral education are negated, as if Goethe's way of science presents some kind of alternative, a counter-culture to the status quo. The aim, rather, is to illuminate the tendencies of our time and to point to a pathway for the metamorphosis of the university in which the achievements of the doctoral style of education are lifted to another level of thinking.

No matter how things are twisted or stretched, it is simply not possible to relate the methodology of conventional objective science in itself to the notion of care. The fruits of this science—for example, medical technologies—may be used for caring purposes but the aim of scientific research itself does not fall within this rubric. Science, conventionally understood, is simply about the dispassionate search for knowledge; what *can* be known *should* be known so that any other way would seem like a form of censorship. This notion of scientific mission is truly beyond good and evil in its methodology; the area of knowledge one elects to work in can of course be motivated by concern for the world but the *technique* of science in any field, be it physics, psychology or biology, is always the same and is objective—that is, 'value free'.

As will be seen, Goethean science, in its techniques, in the whole way it seeks to engage with the phenomena of nature and human society, is an expression of care. The specific example presented in Part III of this book relates to the study of economics and demonstrates the method of study in which we enter into the phenomena of money and capital in an exact, imaginative way and thereby are able to inwardly participate in metamorphic processes in the economic realm. Care gives us the incentive to extend and deepen our understanding because it wishes for things to appear most fully what they are and potentially can be, not diminished or tarnished through any preconception or prejudice. A consciousness of this caringness is the beginning of what it means to teach Goethean science and a teacher needs to work from the very depths of the care which is expressed in the techniques of this science. Young people entering universities today care deeply about the world because it is natural and healthy for human beings to love the things of the world they are born into. This, for the teacher working with the student, is the germinal point in the process of education; it is this attitude in students which needs to be recognized, embraced and enhanced to the highest degree. Only from this point of engagement will anything connected with this way of science ultimately become fruitful in relation to higher learning and human society more broadly.

Care is not the same as concern.[9] Science as a whole is motivated by concern in the sense of interest in and responsibility for the truth; in the Danish philosopher Søren Kierkegaard's account of existential concern it becomes evident that in so-called disinterested objective reflection on things there is, in fact, at work a personal concern with truth. Concern in the normal scientific sense is about the rigorousness of the practice of science and has nothing intrinsically to do with a personal relationship to the thing being studied. Indeed, according to the technique of objective science, one needs to be concerned *not* to develop any kind of personal relationship for this may influence the results of the work by slanting them in the direction of a personal predilection. Goethe's way of science is in no sense a 'lapsing into subjectivity'; on the contrary, the insights it develops from inward, imaginative participation in living form are built on exact observation.

Concern is a word descriptive of a human attitude; care, by contrast, has fundamentally to do with what it means to be a human being. As the philosopher Martin Heidegger has shown, care (*Sorge*) is the very structure of the human self (*Dasein*), the way humans participate and involve themselves in the world.[10] Care is always about a relationship; it is what stirs the will to best serve the subject of one's care. We might say that it is natural for teachers to care about their students and genuine caring relationships can happen at university level when the size and style of the institution permits. But care is not always expressed directly and personally; it can have a powerful avenue of expression in the way teachers guide their students into a relationship with the world around them. The methods and techniques of Goethean scientific research are intrinsically caring in their approach to the study of nature, the human being and human society. The philosopher and artist Rudolf Steiner (the first editor of Goethe's scientific works) comments:

> [Goethe's] intention was to allow the forces inherent in outer objects to come alive in his conceptions of them. This approach to knowledge aspires to an extremely intimate sharing of experience with outer living things.[11]

However, techniques and methods which relate to this approach are not simply there at hand in our culture today. Out of the germinal forms of observation and research we now call Goethean science new ways will need to be developed in relation to all the disciplines.[12]

Goethe and his colleagues were motivated in their work by a profound sense of care for the natural world—different yet anticipatory of what we might today refer to as 'environmental consciousness'. One particular work of scientific thought, probably more than any other, affected these thinkers in a negative way and stimulated their collective sense of concern and care. This was Baron d'Holbach's *Système de la Nature*, which caused a sensation across Europe when it first appeared in 1770. In his *Système* d'Holbach denied the existence of a deity and pictured the universe as nothing more than matter in motion, working inexorably according to the laws of cause and effect. The reaction of many thinkers and artists to this work is encapsulated in the words of Goethe:

> We did not understand how such a book could be so dangerous. It appeared to us so dark, so Cimmerian, so deathlike, that we found it a trouble to endure

> its presence, and shuddered at it as at a spectre . . . A system of nature was
> announced; and therefore we hoped to learn really something of nature, our
> Idol . . . But how hollow and empty did we feel in this melancholy, atheistical
> half-night, in which earth vanished with all its images, heaven with all its stars.[13]

What Goethe felt vanishing as he read this book was nature's inexpressible mysteries, its sublimity, the qualities he had in mind when he penned the words: 'Nature! We are surrounded and embraced by her—powerless to leave her and powerless to enter into her more deeply'.[14] These were the qualities which, for Goethe and his colleagues, made nature worthy of their reverence and deepest solicitude in the way they approached and studied it.

A plant or an animal stands before us; in its simple presence or existence it is yielding itself towards us. Merely this self-yielding presence evokes in us a certain kind of solicitude—the care which also goes by the name of love. This is the wish to preserve what is loved and valued, the wish for it to realize itself to the greatest degree possible within the conditions and potentialities of its existence. It is compassion for the simple fact of its existence, not just because it may be needy in some way, or because it is *doing* anything in particular which may warrant such solicitude. We share simple existence with everything else on this level—the difference is that we are conscious of it in a way which animals and plants are not. For that reason, a responsibility accrues to us, and part of that responsibility relates to *how* we go about gaining knowledge of the living world. In the sense that Goethe meant it, we can say that all the beings of nature are our subjects and deserve our greatest possible solicitude.

During the heyday of German 'nature philosophy' centred around the university at Jena, Goethe was intimately associated with the philosopher Friedrich Schelling and the two of them shared many concerns and interests in relation to the practice of natural science and the arts. In later years Schelling wrote:

> With each explanation this is first and foremost: do justice to that which is to
> be explained; do not suppress it, interpret it away, belittle or maim it in order
> to make it easier to understand. The question here is not: 'What view of the
> phenomenon must we arrive at in order to explain it in accordance with one
> or another philosophy?' Rather, it is the other way around: 'What philosophy
> is required if we are to live up to the object, to be on a level with it?' It is not a
> question of how the phenomenon must be turned, twisted, narrowed or crippled so as to become explicable at all costs on grounds that we have resolved
> once and for all not to go beyond. Rather, to what point must we enlarge our
> thought so that it is in proportion to the phenomenon?[15]

Schelling's poignant philosophical statement points precisely to what is striven for in the practice of Goethe's way of science. It begins with solicitude for the phenomenon as encountered in a physical sense— through exact, unprejudiced observation—and proceeds through an ever deepening mental and imaginative participation in the more 'hidden' levels of its being. It is only in the light of such a respectful, open-hearted and open-minded encounter that the 'open secrets' (as Goethe put it) of things may yield themselves to the researcher.

We can, for example, ask the following questions of the phenomenon of colour: is there a way of understanding a colour which honours and stays true to our sensory experience of colour, to the naked existence of the colour as it yields itself towards us? If we immediately ask *what* a colour is and pursue that path narrowly, we go the way of Enlightenment rationality and the modern wave theory of light by seeking the answer only in an explanation— a law (the wave theory) into which all colour experience is subsumed. In that process colours are turned into numbers—but numbers are not colours! The *quality* of violet, which means the phenomenality of the colour violet, has nothing to do with the fact that its wavelength is four-tenths of a millionth of a metre.[16] Of course the debate about the value and relationship of the two approaches must be had but the most important thing, from the point of view of teaching and practicing Goethean science, is the will to expand our thinking so that it is in proportion to the actual phenomenality of colour.[17] The Goethean science of colour as it may become part of an orientation course of studies in the new university is entered into in Chapter 5.

Care demands that we do not lose sight of the phenomenality of the phenomenon—and it is just here that we may first come to appreciate the significance of Goethe's way of science. Teachers of Goethean science must be prepared to address themselves to the kind of student who is drawn by life's quest and solicitude for things towards the threshold of the sphere of life. This science has no relevance to the kind of tertiary education which puts its emphasis on theoretical thinking and mastery in a material sense. The metamorphosis of the university is about the challenge of allowing itself to be reshaped by a new life impulse. The impulse behind this metamorphosis is the recognition that the doctoral principle cannot guide students to a truly *living* understanding of the world, that it must therefore be transformed—consciously and with great care—so that a new educational ideal can become active within and through it.

The Quest and The Threshold

Goethe depicted the movement of human conscience towards 'care-filled truth' in his *Faust*, the work which occupied him from his twenties until his eighty-second year, so that he was able to pour into it the distillations of his entire spiritual development. Key moments in his own biography found expression in this drama in one way or another, and of particular importance for our theme of education is his famous two-year trip in 1786 to Italy, undertaken at the age of thirty-seven. Coming after already completing the first part of *Faust*, the questions he had brought forth in the writing hovered before him and inspired his adventure in learning.

Goethe's decision to undertake this journey was in response to a crisis in his life. For over ten years he had been involved in largely administrative tasks in the employ of the Duke of Saxe-Weimar and he was desperate to break free. As is often the case, we set forth on new pathways in life with only vague notions of what we can hope to achieve. Goethe's educational experiences had always been unsatisfactory; in his late teens, studying law by rote at the University of Leipzig, his life was riotous and rebellious. Later, at Strasbourg University, he still wasn't happy with the narrow form of legal studies and attended numerous other lectures on scientific subjects. Outside the context of the university he made the acquaintance of the philosopher Johann Herder who greatly stimulated his interest in art. It was now that his interest in Shakespeare and the eighteenth-century Swedish botanist Carl Linnaeus was awakened, two authors from whom he later claimed he had benefited the most. But it was only during his stay in Rome that he had what could be called his first educationally enlightening experience with a cohort of other 'aspirants'.

For the teacher and practitioner of Goethean science, this journey to Italy is most instructive—for indeed, Goethe was learning Goethean science *himself* through his journey. Already as a young man he had come to sense the threshold which stands between intellectual knowledge and a true science of living form and this came about through events and conditions which were particular to his biography. As we shall see in Chapter 2, the requirement to 'live up' to the phenomena of life, to discover how it is possible to be 'on a level' with them, is what is asked of anyone who would enter the sphere of living form. In the teaching of Goethean science, the central question today and for the future is: what is required of us as teachers if we are to bring our students to the experience of this threshold? It is of course not a matter of reproducing Goethe's historical trip but it is of great value to identify the key situations and experiences through which Goethean science came into being.

Firstly, on a general historical note, Goethe lived in a period of European civilization when people sought powerful experiences of threshold. In a certain sense, the meaning of threshold actually defines this era; the French Revolution gave rise to a joyful sense (at least initially) that Europe was on the verge of a totally new, liberated civilization. For many at that time the yearning to break out of the confines of traditional monarchical Christian culture was immense. This was the period of European history when the far reaches of the globe were being explored and opened up for the first time; voyages of exploration were being made to the antipodes and exotic lands such as India were being colonized. Goethe certainly was open to such influences; among the circle in Jena with whom he was intimately involved were the Schlegel brothers who, at the end of the eighteenth century, were the first to publish the Bhagavad Gita and Ramayana in a European language and to develop the study of Indian philology. After 1778 Goethe was in active correspondence with Johann Georg Forster, a German travel-writer, ethnologist and painter who had been on Captain James Cook's second voyage around the world from 1772 to 1775, reaching New Zealand. Among philosophers and artists of this period a certain inner striving found expression in relation to this outward questing; it was the wish to bring the human spirit within the compass of a totally new experience and conception of life.

The physical journey south was the outer aspect of Goethe's quest. What he desired was a transformation in his *thinking* and this had to do with his search for a true organic science, for a thinking commensurate with the living. It was not so much the theoretical deliberations which had surrounded him on all sides in his home situation which inspired him on this quest but the wish for an actual experience of what lay beyond the confines of intellectualism. Already he had been stimulated to transform his thinking through his study of Linneaus; the Linnaean method of classifying and ordering living things was for Goethe a way of thinking which held things apart in a superficial way and did not find their inner, living relationship.[18] Kantianism, the most powerfully influential philosophical outlook of his time, also prepared him for this quest in a significant way. Kant's view was that the human mind is fundamentally divorced from the essential nature of life, from 'things-in-themselves'; he argued that thinking merely reflects the world but can not actually encounter it directly. Kant had hinted at another form of thinking, an *intellectus archetypus*, through which the mind could unite with things and think *out of* their living creative essence, but suggested this was not in fact a human possibility. Here was a philosopher who expounded a deep understanding of organic form while at the same time claiming that the capacity for a 'living' form of thinking was not allotted to the human being. Where Kant perceived an impossibility Goethe beheld a threshold.[19]

Even in the early part of his journey, when passing from the Germanic northern lands over the Alps into Italy, with Rome and its remains of Classical culture as his much-desired destination, the mountain environment provided for him a certain threshold experience—a threshold in seeing. After observing that the form of the same plant species is different in the heights from on the plains, he penned in his diary:

> The truth is that, in putting my powers of observations to the test, I have found a new interest in life. How far will my scientific and general knowledge take me? Can I learn to look at things with clear, fresh eyes? How much

can I take in at a single glance? Can the grooves of old mental habits be effaced? This is what I am trying to discover.[20]

Here is the first principle or element of the phenomenological process of observation we now call Goethe's way of science. It is the dedicated effort to perceive the phenomenon exactly, in terms of itself, rather than through what our viewing may impose upon it. For the teacher the task is to orientate the students in a fundamental way to the intentions of this science. There are many ways by which such an orientation can take place but the main point is this: unless every starting-out, in every observation process—be it a mineral, a plant, animal or any human phenomenon which is being observed—has a sense of quest, of entering into the mystery which the outer appearance of the phenomenon only indicates, then this science is not actually taking place and no threshold will ever be reached. Only such a reverential disposition can provide the inspiration and impetus to efface one's mental habits and clear one's vision for the sake of another being.

Later, in Rome mainly, Goethe had the opportunity to engage in an extended study of works of Classical art—including the Belvedere Apollo—with the group of like-minded individuals he soon came in contact with. This was no touristic artifying but an activity of intense collaborative investigation. The group was always visiting the galleries and moving through architectural settings with their notebooks and sketchbooks. He wrote:

I am very happy here. All day and far into the night we draw, paint, sketch in inks and practise the arts and crafts very much *ex professo*.[21]

Fig. 1: The Belvedere Apollo, a sculpture from classical antiquity studied by Goethe in Rome.

Aside from the great intensity of this investigative practice, here we can identify a second element in the Goethean phenomenological method. It is clear from his description that the way he went about working was not *only* discursive—that is, not just theoretical debate and philosophizing—but always involved artistic practice of some form. The reference here is to the study of works of art but the same thing applies to the phenomena of nature when we are considering Goethean science. Wherever he went on his travels, whatever he was observing, his observations always involved visual artistic and poetic reflection. Artistic practice is one of the ways in which all observations become an activity of the whole human being—we could say, not just of the head but also of the heart and hands.

Further in relation to the study of art, Goethe makes several statements in his diary while in Rome which reveal how closely he associated art and nature. Through penetrating and repeated studies of master artworks he came to the realization that these works were created with the same formative laws which give rise to the works of nature. He writes: 'I believe I am on the track of these [laws].'[22] If one makes a serious attempt to reckon with what Goethe means here, the intellectual mind reaches a threshold—inevitably. The idea challenges, even disrupts, the conventional way of thinking and prepares it for the new—for it fits within no conventional frame of reference to suggest that nature and art are united by the same lawfulness. The status quo view is that the processes of nature and artistic production are activities of entirely different kinds, the first objective, factual and the second of a subjective, personal nature.

Students of Goethean science need to be guided towards the possibility of experiencing for themselves what Goethe came to regarding the relationship of art and nature. Philosophical reflection is not enough; it is only an initial stage. The threshold to a true science of organism is approached but not crossed by philosophy. Goethe was not content to theorize about the relationship of art and nature—he wanted to *perceive* the lawfulness which unites them, to actually experience and work within it. Thinking which has lifted itself to the level of life has metamorphosed itself into the power of dynamic, pictorial cognition which is called *cognitive imagination*; artistic perception thus must become an organ of scientific cognition. If the creative process of art is the same as nature, then the experience and practice of art—art being the immediate product of the human spirit—becomes the doorway through which the human spirit gains entrance to an understanding of the creative laws of nature.

More or less consciously, through his studies in Rome, Goethe was preparing himself for the threshold moment in cognition when the seeker of knowledge is effaced, when the mind conditions itself in order to receive into itself the form of the thing it is endeavouring to understand. Such a moment occurred for him at the farthest point of his journey in the botanical

gardens of Palermo in Sicily, where he writes that a 'spirit' seized him in
a condition of poetic meditation and he understood the archetypal nature
of plants.[23] This experience became the key to the development of his biological studies in the following years. The crucial fact is that he reached
this point of biological insight through an apprenticeship in the art galleries of Rome. Art shapes the organs of imaginative perception which then,
if directed and cultivated in particular ways, become faculties of natural
scientific research.

Goethe's apprenticeship in Rome was not carried out in a solitary manner—we have already noted his close involvement with a group of artists there. Something about this collective art-making experience greatly
intensified in Goethe a way of understanding which he was unable to
make fruitful prior to his Italian journey. He himself provided a clue
among his diary entries of this time:

> In the artistic colony one lives, as it were, in a room full of mirrors where, whether
> you like it or not, you keep seeing yourself and others over and over again.[24]

What is evident here is the extraordinary *engagement* of his learning experience; such learning is a very far cry from the impersonal and intellectually
detached mood which typifies the lecture and seminar experiences provided by doctoral education. In such an intimate social context, enlivened in
particular through shared artistic experience, a deep sense of mutual quest
can arise in which the gaining of knowledge of oneself and knowledge of the
world cannot be separated. Such intimacy between people is not extraneous
to the process of learning Goethean science; indeed, it can be an important
condition for it. The capacity for Goethean science grows in such a situation because a student is learning the kind of devoted receptivity which is
required in relation to *all* living beings. The quality of receptive listening and
speaking which has come to be known as 'Goethean conversation' is further
explored in Chapter 4 in relation to the teaching of Goethean science.

Goethe set out on his journey to Italy impelled by a sense of quest, a
quest for a new way of understanding the worlds of culture and nature.
Retrospectively we can see that the journey had a certain shaping; through
particular events and experiences he was guided towards the threshold of
the living whole and the possibility of a true life science. Goethe's quest is
not necessarily to be imitated but in its overall character as an educational
experience it offers much which can be of assistance to us on the difficult pathway beyond doctoral education. Each of its main features—the
personal experience of quest and threshold, the relationship of art and
science in life science, the social context of learning—are explored on the
following pages.

Chapter 2
Guidance Into The Sphere of The Living Whole

The Question of Life and Cognitive Feeling

We wish to guide our tertiary students into the sphere of the living whole—that is, into the practice of a true life science. This is the challenge of the teacher of Goethean science: to create a teaching which involves a living, intimate engagement with the forms of nature and human society, more than just the tossing and turning of concepts and theories. At the most fundamental level the students need to experience the difference between inorganic and organic science, for without this understanding nothing pertaining to Goethean science will have a sense of value and imperative.

At the entrance to the realm of life stands a 'guardian' in the form of an obligation issued forth in the name of the totality of beings which grace the planet Earth. This obligation is not to be underestimated. To pass as a teacher, student or researcher into the sphere of the living whole there must be *a metamorphosis of thinking* and this is demanded, not by any human agent, not by any set of rules or moral codes, but by the living beings which we seek to understand. The living phenomenon, if rightly seen and valued, 'speaks' to the conscience of the student or researcher who comes to realize that no step forward into a true life science can be made through just the exercise of the logical intellect. The whole human being must become active, is *called on* to become active, in order to comprehend the living world and its silent demand.

Conventional science experiences no threshold, hears no demand, when carrying out its investigations of living things; hence it uses the same thinking it has developed in relation to the inorganic and applies it to the organic. For this science the passage from the inorganic to the organic appears to be uninterrupted, seamless. This has to do with the history and philosophy of science; science formed its hypothetico-deductive methodology in relation to the elucidation of mechanism, through a comprehension of cause and effect relations. It was mechanics (inorganic science) which rose to prominence during the eighteenth-century Enlightenment, a notable example being the 'celestial mechanics' of Isaac Newton. From the time of the Enlightenment into our own this methodology has become an immensely powerful tool for researching every aspect of the physical world. The biological sciences (and later sociology, psychology, political science and so on) took form by applying and adapting the kind of methodology which had first been refined in the realm of mechanics.

In actual fact, the conventional sciences do *not* pass into the sphere of the living because they do not accommodate themselves to the living through their methodology. This is the state of affairs which needs finally to be reckoned with at the beginning of the twenty-first century, when our immense technological confidence must be connected with the immense problems we have created in relation to the life sphere of the planet. Science takes hold of living beings but sees them in a very particular way which has to do with its procedures; it perceives living form through a mechanical frame of reference and searches only for mechanisms which it then adds to its store of 'truths'. Friedrich Nietzsche had a sense of this when he wrote:

> Should one lay hold of [scholars], then do they raise a dust like flour-sacks, and involuntarily: but who would divine that their dust came from corn, and from the yellow delight of the summer fields?[25]

We must consider the fact that it is possible to have a living phenomenon right under our noses as it were, to cut into it with our methodological scalpel, yet even so remain very far distant from the sphere of life in our thinking.

Therefore, the very beginning of the teaching of Goethe's science involves the awakening of the faculty of living thinking, the thinking which thinks *in* things, which cognitively experiences them, which abides within the yellow delight of the summer fields and does not merely generate dust. This can be initiated in a relatively simple way; let us contemplate the colours blue and yellow (see Fig. 2).

To practice this involved thinking we mentally 'enter' each colour, we experience the colour as a phenomenon with a very definite quality or inner necessity. We should be able to say, as Goethe did, that 'my thinking is not separate from objects . . .', that our thinking is in the object, but equally we could say that the object is in our thinking.[26] This is most certainly *thinking* but it is not an intellectual-logical thinking; it is actually an intimate *encounter* with the colour, not just an articulation of thoughts about colour. Now, we must ask ourselves: what aspect of ourselves comes into play here? We feel the precise quality or inner activity of the colour in a way which is replete with comprehension; blue retreats into itself; yellow radiates out in all directions. This is what we call a *cognitive feeling*. At the moment we first experience cognitive feeling we have taken one step beyond the threshold of the sphere of living form.

We can compare this experience in thinking with a simple mental process of a strictly logical kind. Let us consider the following mathematical equation:

$$8 + 8 = 16$$

Fig. 2

Fig. 3: Kazimir Malevich, Abstract Composition, oil on canvas, 1915.

As we carry out this operation in our minds with a heightened awareness of what is taking place, we find that the necessity of the sequence
doesn't depend on anything outside itself. Logically this process is
self-contained or self-sufficient. We do not experience ourselves to be
involved in this operation; we simply consider it and grasp that it is
necessarily true.

Conventional science rests upon the self-evident mathematical logic which is exemplified in this equation and precisely the same kind of logic is applied to the study of organisms. In every case exactly the same kind of 'solid' definiteness is sought as the basis of scientific truthfulness. The question is—and it is a very large question—to what extent is the organism being squeezed into the terms of our logical thought operations?

In other words, are we forcing the organism to 'obey' the self-sufficient logic of our mental operation rather than seeking to raise our thinking to the level of the organism? If an organism such as a plant or animal is studied in a piecemeal fashion, logic certainly holds good in relation to particular aspects or processes. But do we ever catch sight of the whole organism through such mental activity? In Goethe's way of science the whole is sought at every moment and this quality of wholeness is not the product of a logical thought operation but is encountered or experienced in the particulars through cognitive feeling.

Much has now been written about Goethe's science, both from the philosophical and practical points of view. However, certain statements stand out as particularly cogent and fundamental. On a few pages in his book *The Science of Knowing* Rudolf Steiner outlines his reasons for claiming that Goethe was the initiator of a true science of organism. To guide students into the sphere of the living whole two especially valuable sets of statements from this text are set out below and developed through artistic examples. Steiner's text was written around a hundred years ago but today it has something extraordinarily fresh and contemporary about it, as if just now ripe for being understood.

The Comparative Method

Steiner, in this text, makes the following points about method:

> . . . the erroneous view [is] that the method of a science is extraneous to its objects of study, that it is not determined by these objects but rather by *our* own nature. It is believed that one must think in a particular way about objects, that one must indeed think about all objects—throughout the entire universe—in the same way . . .
> . . . The ideal [of organic science] will be fulfilled when *the comparative method employed by Goethe* is recognized in all its implications.[27]

Building on our 'elementary' experiences of cognitive feeling in relation to the two colours we can now explore how these experiences can be heightened; in this we can follow the indications of Goethe himself. We have seen in the previous chapter how he derived profound insights into the creative laws of nature on his Italian journey from the contemplation of great works of art. Entering the sphere of the living whole, we are obliged to part company with the view that science is a realm entirely divorced from that of art.

Goethe occupied himself with masterpieces of Classical figurative art, but in fact modern abstract art is not 'burdened' by the close representation of physical reality and brings the laws of artistic creation to the surface as it were, making them visible. With the guidance of such art it is possible to look directly into the realm of organic formation and the laws of the organic whole. For some modern artists this has been an entirely conscious and intentional act and certainly Kandinsky strove in this direction. For many modern abstract artists, painting approaches music by virtue of the fact that, through colour, line and shape, it expresses a language of gesture. Such works may serve as guides in teaching the Goethean science of living form.

We may cast our eyes over a work such as *Abstract Composition* by Kazimir Malevich (see Fig. 3). What shines out is the fact that it is composed of many parts and these parts interrelate unlike anything else we encounter in an immediate way in the physical world. This painting is not just another 'thing' amongst the things we find around us; the work opens a window upon another world of 'hidden' relationships. This sets us off balance somewhat, disturbs our complacency and feeling of familiarity towards things. Just this experience, from the point of view of teaching, can be powerfully instructive along the path of an organic science. All of a sudden we are not the masters of the situation, determining the precise conditions of the experiment. We are put in a receptive mood with our thinking and feeling; as Pablo Picasso once commented: 'I do not seek, I find.'[28]

The degree of openness of our approach determines the degree of revelation of the work we are contemplating; we gradually reach the point where it becomes possible to gain insight into its formative idea. Every choice of colour and form, every relationship of every part, is dictated by the whole in a genuine, masterful creative work. The creation of a work of art is an organic process in the sense that it comes into being through a living creative activity; the work is of course not physically alive. In an artwork like *Abstract Composition* only relationships, gestures and formative forces shine out; there are no references to trees, people, buildings and so on. The aim of the teacher is to guide the student beyond a certain threshold which is the point where one loses interest in one's own preconceptions or opinions and becomes devoted to finding the work's formative law.

We employ the 'comparative method' to understand this work by Malevich in its totality. Because of its abstract nature we easily find ourselves seeking to understand the work by sensing the relationship of each element to every other element—the horizontal black and red rectangles on the right which appear as polar to the black half-circle on the left, the sloping shapes below the central red shape and the thin sloping background lines. This sensing of relationships between the elements of the whole is an exact process but it is not logical; just as with our experience of the colours yellow and blue, we mentally 'enter' and participate in the work and *experience* its inner dynamic or gestural structure. Patterns of form emerge which reveal one part's necessity in terms of every other part. Through cognitive feeling we are able to participate in each shape and colour in its exactness and sense every relationship so that gradually we come to experience the lawfulness of the work as a wholeness.

Working with art in this way is a step on the path towards the awakening of the faculty of cognitive feeling or what Goethe called 'exact sensory imagination'. The important thing is that, according to Goethe, this same faculty must be made active in order to understand living form. The organism, too, is composed of parts in dynamic relationships which, as Kant made clear so long ago, are mutually creative or self-referential.[29] The organism expresses a wholeness or what Goethe called an 'inner completeness'. In the case of the work of art the essence or lawfulness (necessity) of the form is revealed and shines forth because the work is an expression of the human creative spirit (Goethe called this the 'spiritual-organic'); in the case of the organism the lawfulness is hidden, as it were, within the physical manifestation of the organism and has to be brought to light through study.

So when we turn to a contemplation of an organism such as the giraffe, we proceed in just the same way as with the work of art; step by step we build a rich inner picture of this organism without any immediate requirement to analyse or interpret. We observe, we note in a factual way, we draw upon the empirical work of others. Each piece of empirical information is entered into with cognitive feeling and inwardly sensed and 'tested' in relation to every other. In other words, we carry out the 'comparative method'. Properly, carefully, respectfully observed, each and every part of this animal dictates exactly how it is to be experienced. The form and colour of the body, the shapes of the bones and their connections, the features of the musculature, the relative sizes of the parts, the principal behavioural characteristics, the cry, even the way the animal dies—all such observations of form and function are carefully, systematically, compared with each other and brought together build a comprehensive imagination of this animal. Writes the Goethean scientist Craig Holdrege: 'With the iron will to always return to the whole and the discipline to hold back from short-cut explanations, I keep the path to the unity of the organism open.'[30] Gradually, as with our study of the work of art, we come to a cognitive experience of the organism.

A prominent character of the giraffe is of course its long neck. When taken in isolation short-cut explanations can quickly press forward which point to the advantage the giraffe has over other animals which cannot reach high-up acacia leaves. Entering into the sphere of the wholeness of the giraffe through the practice of cognitive feeling, patterns begin to emerge and the long neck becomes meaningful in terms of the whole. In the words of Holdrege:

> When we study [the giraffe's neck] within the context of the whole animal, we discover an overriding tendency toward vertical lengthening that comes to fullest expression in the neck. This tendency shows itself in the legs, in the head, and in the shortening of the body. It becomes a key to understanding much that is unusual and special about the giraffe: the way it stands up, walks, and runs; the way it can reach so high; the way it awkwardly spreads its legs to drink; the way males spar with their rhythmically batting necks; its sensory focus in the overviewing eye.[31]

Holdrege details many other features which relate to vertical lengthening in the giraffe including the size and position of the heart and the structure of the arteries. Indeed, *every* aspect of this animal bears a relationship to its dominant verticality. Just as patterns of similarity and polarity emerge in the work of abstract art by Kazimir Malevich—relationships and patterns which unite the central sloping red shape with every other element of the composition, curved or linear, sloping or horizontal, red, black, yellow ochre, blue or magenta—so it is with the organism. Through an imaginative thinking which carefully and caringly allows these patterns to emerge and coalesce, the idea of an organism gradually comes to light and in our thinking we come to a comprehension of its living wholeness.

The Typus

Rudolf Steiner says the following concerning the lawfulness of organic phenomena:

> The *typus* plays the same role in the organic world as natural law does in the inorganic . . .
>
> . . . A law governs the phenomenon as something standing over it; the *typus* flows into the individual living being; it identifies itself with it . . .
>
> . . . The [*typus*] determines only the lawfulness of its own parts. It does not point, like a natural law, beyond itself. . .
>
> . . . Every single organism is the development of the *typus* into a particular form. Every organism is an individuality that governs and determines itself from a centre. It is a self-enclosed whole, which in inorganic nature is only the case with the *cosmos*. . .[32]

Why must we concern ourselves with this *typus*? The *typus* belongs to the sphere of organic wholeness; it is incomprehensible outside this sphere. The *typus* is recognizable to a living, imaginative form of thinking and although philosophical considerations may guide us towards it, it cannot be experienced by the logical mind. It is another name for the wholeness of the organism, and is the name which Goethe used. As developed here, for educational purposes we orientate our students towards an understanding of the *typus* through artistic practice.

A natural law is something which is founded in our experience of material existence. We come to know that our body and every other physical object around us interact with each other in a cause and effect manner. If we cut the trunk of a tree it will predictably fall; if we kick a ball it will predictably rebound from our foot. The logic and predictability of material existence comes to expression in the simple equation $8 + 8 = 16$ in which we recognize the lawfulness of material relations. We may take eight of anything in the world—shells, stones, people—and the same addition will apply; the qualities of each shell, stone or person are irrelevant. Just so with Newton's second law of motion: $F = ma$ (force = mass X acceleration)—it doesn't matter what physical entity it is applied to (it could be a stone or it could be a bird, or any physical entity in the whole universe) the acceleration of this entity will obey the same law. The lawfulness of these two equations stands 'above' or 'outside' any particular thing in the universe and governs all things equally.

If our perspective is outside the sphere of the living whole we will take the view that a phenomenon such as biological evolution must also follow a natural law—and such a law will be sought. The same logical steps will be taken; just as $8 + 8 = 16$ stands 'above' any particular objects, so too the law of biological evolution is thought to stand above any particular natural phenomenon. The Darwinian theory of natural selection is considered to fulfil the requirements of such a natural law. This theory follows a stepwise logic which is self-sufficient and as logically coherent as our simple equation:

a. Differences or variations occur among individuals of a species.
b. Some variations are helpful, and individuals with helpful variations survive and reproduce.
c. Organisms produce more offspring than can survive.
d. Over time offspring with helpful variations make more of a population and can become a separate species.

This mechanism 'points beyond itself' to every particular instance of evolution. Again, the qualities or characteristics of the instance are irrelevant; it applies to any species, plant or animal, and governs each equally. While the object of our interest is here with living phenomena the important thing to note is that the Darwinian law of natural selection is not a *typus* and natural selection is not a 'living idea'. With natural selection we are merely treating living things in the logical terms of the inorganic, without attempting to raise our thinking to the level of the organic.

An understanding of the *typus* can be cultivated through artistically-shaped observational tasks. Students will need to accommodate themselves to the idea that, on entering the sphere of living form, artistic creativity becomes a primary way of proceeding. They need to come to a clear realization that the *typus* is not just a concept among other concepts but a *creative activity*. This takes us into scientifically unfamiliar territory—and rightly so, for we are not attempting to bring down the organism to the level of our conventional scientific thinking. Artistic practice is our guide across the threshold into the sphere of the living whole; it makes us familiar with the formative activity of the *typus*. It does this because *art is of the same creative order as the typus*, it is commensurate with it. We must creatively and comprehendingly 'do' the *typus*, enter into its creative process—or, no matter how sophisticated our philosophizing, the *typus* will not be adequately understood. Ultimately what is being sought is not art as such but the artistic sensibility raised to a higher possibility of itself, becoming the faculties of understanding called *cognitive feeling* and *cognitive will*.

A preliminary task could be in comparative anatomy; here we may look at the five-digit limb form (the so-called pentadactyl limb) which is found in various animal groups, including mammals, birds and reptiles. Three mammals have been selected for study here. Working comparatively with the skeletal form of the limbs of these animals is an opportunity to develop a sense of the creative sculptural activity of the *typus* in an exact cognitive way.

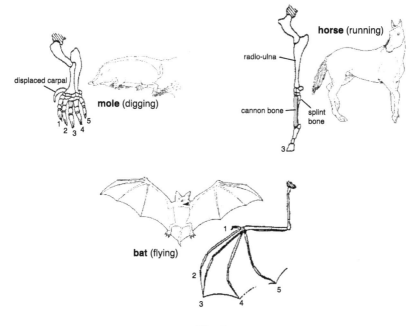

Fig. 4

The aim is to inwardly experience how the five-digit *typus* has 'flowed into' and identified with particular limb forms, governing or determining the shaping of each limb-form as 'from a centre'. Steiner's expression 'from a centre' exactly describes the artistic experience of the activity of the *typus*; it suggests not a spatial but a creative orientation. Artistically (that is, through cognitive feeling) we do what the formative *typus* does: we flow into and identify with each part of each limb form—*exactly*. We re-enact the formative action of the *typus*. Mentally we shape each element of the limb just as we shape the elements of a work of art into a balanced wholeness. With cognitive feeling we do what our hands are doing when they are modelling clay—contracting and squeezing, elongating, expanding and thinning and so on. We may speak of a 'sculptural thinking' when we are thinking *within* the phenomena as a sculptural modelling process, recreating it in our cognitive imagination.

Let us 'think with' the creative pentadactyl *typus* in relation to the limb of the mole. With the mole we must imagine each bone, and the entire limb complex, shortening and becoming thick and robust. The four robust digits spread out into a spade shape (compared to the giraffe which maintains a singular elongated and narrow shaping); the bone of the upper limb of the mole (the humerus) is flattened and outspread. We experience the *typus* as something plastic whereby whatever happens in one bone (say, the bones which makes up one digit of the mole) must relate to what happens in all other bones (which, in a larger study, would include the bones of the whole skeleton). For, indeed, the mole is a compact, robust animal altogether.

This creative scientific work becomes more meaningful when we include a context, because animals are unified with or perfectly adapted to their native terrestrial environment. In the case of the mole we must imagine the formative activity of the *typus* in a kind of conversation with the subterranean environment; we mentally shape the limb of the mole in relation to our mental picture of specific conditions below the ground in which the burrowing function is all-important. This is the specific lawfulness of the animal we call the mole—the lawfulness of the whole animal. In our cognitive, sculptural imagination we are re-enacting the formative activity of the *typus* in relation to the subterranean environment in which the mole powerfully burrows and scoops.

Similar sculptural explorations of the limb bones of the horse and bat lead to the experience in cognitive feeling that the limb bones of these animals are shaped by the same formative impulse of the pentadactyl *typus*. Working with the horse limb in our sculptural imagination we experience a relative lengthening, thinning and straightening of the entire limb such that it stands on the tip of just one digit (the other four digits being reduced to boney lumps). We inwardly model this limb in relation to its original habitat, the hard surfaces of the north American plains and the steppes of Eurasia on which it could run at great speed. With the bat we radically thin, elongate and separate or 'refine' the digits to form a wing which carries the animal through the airy realm.

Working out of such exercises in sculptural thinking we can move to a consideration of evolutionary process. The horse is an organism in relation to which a particularly rich fossil record exists; this record is made of extinct animals which belong to the same type (the artistic reconstruction of these animals from the fossils is shown in Fig. 5). Even to superficial observation these animals are similar to one another or 'belong together'—and a more detailed study would show that the members all share hoofed feet and an odd number of toes on each foot, as well as mobile upper lips and a similar tooth structure. The standard interpretation is that these animals belong together as a developmental sequence, meaning the lowest *turned into* the second and so on. In terms of Darwin's hypothesis this development is what is meant by evolution.

The vital task for the student of Goethean science is to stay within the practice of cognitive imagination, 'with an iron will' keeping open the pathway to the living whole. It is not necessary for students to concern themselves with the so-called evolutionary law—the mechanism of natural selection—as if this is the all-important question. We have already considered that any natural law pertains to the inorganic realm, and our concern in this exercise is to practice a method which is derived from the nature of the living entities themselves. We simply continue to

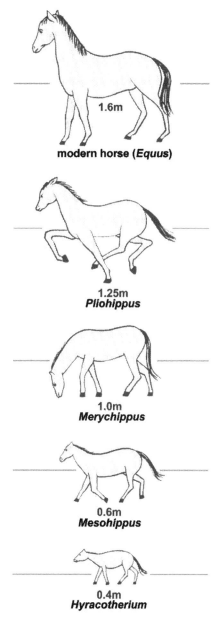

modern horse (Equus)

1.6m

Pliohippus

1.25m

Merychippus

1.0m

Mesohippus

0.6m

Hyracotherium

0.4m

Fig. 5: Horse evolution

work with the comparative method and remain open to whatever insights may flow from it. We work with these horse forms through cognitive feeling and to do this we do not need to *assume* that these horse forms physically turned into each other.

The horse *typus* which belongs to this evolutionary sequence has 'flowed into' each organism and identified with it. We inwardly, dynamically participate in the creative process by 'flowing into' the lowest form in a precise way, the same with the next, and so on through the sequence—also moving backwards. We experience through sculptural thinking how the *typus* has created each organism uniquely, even if similarly. In teaching practice this does not necessarily have to involve an in-depth examination of all the information pertaining to each fossil and its reconstruction; for orientation study purposes just the images above can suffice. We mentally re-enact the formative activity of the *typus* and perceive that *the first animal in the sequence, just as the last and each in between, is the creation of the* typus. Even if the different animals in the sequence have not turned into one another physically, it is nevertheless apparent to the eye of the imagination that the 'idea' of the horse has intensified in a progressive way, from the first to the last. The word 'intensify' is used here in the same sense meant when we speak of metamorphosis in the plant realm, for example in the sequence of leaves present on the same plant at the same time—the lowest being nearest the root and the highest nearest the flower (see Fig. 6).

Fig. 6: Leaf metamorphosis of the field poppy (Papaver rhoeas).

Goethean science has had an extensive exploration in the literature; how and why Goethe went about his research two hundred years ago has been explicated in considerable detail, in both philosophical and practical terms. We have now arrived at the point in the gestation of this science when a further step must be taken, the step which will allow this science to find its proper place in the realm of higher learning. Students of tomorrow will need to be given the opportunity to cultivate the imaginative form of cognition through which Goethean science can actually be experienced and understood. At the threshold to the sphere of the living whole a great responsibility devolves upon institutions of higher education—to carry this life science forward and realize its potential for cultural and social renewal.

Chapter 3
The Metamorphosis of The University

The University At The Threshold

The time is certainly ripe for the development of a tertiary curriculum in Goethean science and this book is written on the basis that it must be taught, that it needs and deserves to find its way out into the world in order to have the salutary impact it potentially can. The question is whether the university as we currently know it is a suitable context for this development, not so much whether the modern university can find a place for Goethean science in its very wide range of courses and programmes. The concern here is with the potential coming into being of a form of university based on a new ideal of education—the ideal of the whole or universal human being—in which Goethean science would have a natural and necessary place.

The institution of the university, like any other complex form of social organization, is fluid to a certain degree and has been able to respond to changing social conditions since its inception. In the last decades many changes have taken place on an 'adapt and survive' basis and the overall trend of these changes can be considered as part of a reflection on the needs of tertiary education in the contemporary world. A particularly heightened moment of self-reflection for universities took place during the 1960s and 70s, inspired in part by the revolts of disaffected students and staff who demanded the university liberate itself from the strictures of tradition. And in a certain respect universities have indeed undergone such a liberating process; one of the key changes of the second half of the twentieth century is that universities have tended to become more diversified and less prescriptive, allowing students freedom in the building up of their courses of study. Rather than a narrow, determined curriculum, what we have seen is the birth of the modern 'multi-versity' with its plethora of courses and research facilities.

Notwithstanding these kinds of structural changes, powerful currents of discontent continue to be active within the university system, views which have sought to challenge the universities in more fundamental ways and whose voices have not been easily stilled. Today there is a whole literature dealing with what ails the modern university, books which could not have been written even 40 or 50 years ago; these include *Education's End, The University in Ruins, Killing the Spirit, The Last Professors, The Closing of the American Mind, The Western University on Trial.*[33] The observations and critiques made in such books are diverse but all seek a reappraisal of the essential values of universities, all reach down to the very 'idea' of the university. Whether the university in its current form can reckon with the following kinds of observations and criticisms remains to be seen:

> An overemphasis on intellectual knowledge. . . has made the university sterile and two-dimensional, depriving it, and human society through it, of the depth dimension that comes from other ways of knowing, especially ways of knowing that would be regarded as instinctive or intuitive or poetic.[34]
>
> It is now very difficult to speak about wisdom in the university, for modern science is not wisdom, rather mostly operational knowledge. If we do not establish a sapiential dimension of academic life, if we do not seek truth that is embedded in wisdom, if we do not seek 'illumination', as the Oxford motto has it, we shall fail.[35]

These critiques have to do, not so much with course content and curriculum structure, but with the *kinds of thinking* which can be cultivated by the universities—and such observations are very important when we contemplate the teaching of Goethean science. With all the political and economic pressures pulling them in different directions together with the force of inertia inherent in extremely large and somewhat unwieldy institutions, it probably stands to reason that change on these levels is extraordinarily difficult and even unlikely. Fresh beginnings will need to be made and new kinds of institutions will need to come into being.

Universities have always been a locus for the ideals and aspirations of their time. They are microcosms, 'little cities', somewhat self-contained and detached from surrounding nature and society and bringing everything from that surrounding world to a kind of heightened consciousness. The university, in other words, is not just a mirror, passively reflecting its surroundings, but is actively engaged with the world through the methods by which it comes to understand and transform it. This points to a key responsibility of the university: its task is not just to utilize and refine the current forms and methodologies of knowledge but to prepare the ground for the coming into being of forms of thinking which are only germinal and still somewhat below the surface of cultural life.

As concerns Goethe's way of science, this has figured in no way at all in current debates on the future of university education, for it is still removed from the general field of contemporary consciousness (although it has increasingly come to feature in tertiary courses on German romanticism, aesthetics and environmental philosophy and the history and philosophy of science).[36] For this reason it may appear as rather unusual to speak about Goethean science in the way which is being done here— that the university system must adapt itself to the kind of education required by Goethe's science of living form, rather than the other way round. But Goethean science is not an isolated cultural movement; it is an expression of a powerful and deep current in Western civilization which rose to singular manifestation around the turn of the nineteenth century. The threshold moment for Goethean science—not yet arrived but inevitable—is the point where it loses its attachment to Goethe the individual and becomes broadly recognized for what it actually is: an authentic science of the living world. Universities will then need to respond to the imperative to shape themselves according to a new form of thinking, not just a new content of knowledge. The movement of tertiary education from inorganic (mechanistic) to a living, imaginative form of thinking is what we are referring to here as the metamorphosis of the university.

The significance of Goethean science in the debate about the universities thus begins to appear when we draw our perception deeper into the currents and forces at work below the surface of civilization, when thinking about the university *itself* becomes a study of a living form—the dynamic, evolving phenomenon called human civilization. The historical development of the university in Europe can be pictured in three main stages of metamorphosis. The institution of higher learning in the Middle Ages was called a *studium generale*; this was a community of teachers and scholars which conducted itself in a formal way by working systematically through classical texts, raising questions and debating points in order to uncover and establish truths in theology and in science. Centuries after the advent of Aristotelian logic, logic was a major subject of study in the first universities and was the method of proceeding in theology as much as in mathematics and geometry.[37] In theology the chief aim of the logical method was proof of the existence of God; God was the unifying principle of the medieval university, considered transcendental and external to the university but the subject of unalloyed attention. God was the binding force of the medieval community of scholars. 'Faith,' it was held, 'precedes science, fixes its boundaries, and prescribes its conditions.'[38] Theology, along with law and medicine, constituted the 'upper faculty' of the medieval university; before that the student passed through a bachelor's degree of the liberal arts which involved the seven subjects of the ancient *trivium* and *quadrivium*—grammar, logic and rhetoric followed by arithmetic, geometry, music (harmonic theory) and astronomy. The *trivium* was the means for finding the way to structure and express knowledge; the *quadrivium* was about the aspects of the study of number as the foundation for universal knowledge.[39]

The modern university was ushered in by Kant with his picture of the unifying principle of reason, the power of intellect beyond the mere application of logic. In the period intervening between the Renaissance and the time of Kant there awoke the sense that in human rationality the divine power itself resided and this was the basis for the coming about of a new 'reborn' humanism. By the time of Kant it could be conceived that reason was the unifying principle *immanent* in the university, uniting all faculties without the dependence on faith in an external God.[40] Consequently the intellectual centre of the university became philosophy rather than theology, and was expounded as such as the basis for the first modern university, the Humboldt University of Berlin, by the philosophers Schelling, Wilhelm von Humboldt, Schleiermacher and Fichte.[41]

A third step brings us to the contemporary period and the great 'multi-versities' of our time, tertiary institutions embracing a multitude of subjects and approaches without any unifying principle and serving mainly pragmatic, socially and economically defined ends through 'exchanges of information' reliant to a large extent on the computer with organizational approaches drawn from the corporate world.[42] Philosophy is no longer the central faculty of the university, reason no longer holds its place as the immanent and unifying principle; postmodern scepticism has subverted reason's value and undermined its status. While we have campuses as functional situations of research and learning, they are no longer cultural-spiritual centres which, in the second phase of development, had been understood as the context in which individuals could attain to self-realization through the exercise of the power of rationality.[43] The idea of the modern university does not relate to ideals of truth and universal meaning but is more about career pathways and providing for individual student interests.

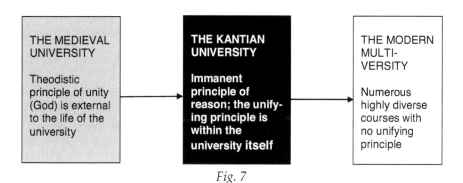

Fig. 7

Here are the three principal steps (Fig. 7), set out in graphic sequential form as stages of a metamorphosis. It is important to work with this goetheanistically—just as with any other metamorphic process in nature and the human world—to enter into it and experience through cognitive feeling what is taking place. We move through the sequence so as to become very intimate with it in our inner life; then it slowly becomes something far more than a series of historical facts. It becomes a mobile living picture of the evolution of human consciousness.

The focus of our concern with Goethean science is firstly in relation to the second stage of this development, for this was the time when Goethe stood as a towering and highly influential figure within Germanic culture. Although Goethe was influential in a general way, the actual practice of his science was not incorporated within the new kind of university which came into existence in his time. Goethean science took shape as a germinal bud, a potentiality, and has remained so while the modern university eventually took a quite different direction. Goethe's science rests on a heightened artistic capacity—a cognitive feeling or exact imagination—and to this extent it belongs to the phase closely connected to but beyond the strictly Kantian idea of the university. Through Friedrich Schelling and other philosophers of the immediate post-Kantian period—and particularly through Wilhelm von Humboldt himself—Goethe had an indirect influence on the founding of the Humboldt University of Berlin. This is what made the founding of this university in 1810 so extraordinarily significant—so much was in the melting pot at that time and so much was at stake.

Friedrich Schelling is noteworthy in this connection. In the summer of 1802 he gave a series of lectures in Jena entitled 'Vorlesungen über die Methode des akademischen Studiums'.[44] In these lectures Schelling makes a case for university study involving a way of thinking which is able to penetrate to the wholeness of life, as opposed to the utilitarian concept of education which he calls 'bread-and-butter studies'. The latter, he suggests, do not enter into the principles behind science but, rather, see science 'purely as an apparatus for achieving material ends'.[45] To grasp a science in its particularities or its mere factuality, he writes, is to be a slave to it. What he is extolling is a way of philosophical thinking about the facts and procedures of science which is imbued with the power of imagination: 'Living interest in scientific knowledge develops the creative imagination, the intuitive vision in which the universal and particular are always one.'[46] Elsewhere he writes that, at the point where knowing becomes a creative, imaginative form of thinking, it is able to cognize the creative forces by which the forms of nature come into existence, the same forces which are at work in the creative intuitions of the artist.[47]

Now it is true that Schelling was concerned with philosophy and not specifically with Goethean science which proceeds through observational or experimental work. But Schelling's philosophy is definitely orientated in the direction which Goethe went in the practice of life science and, as Goethe himself recognized through his close collaboration and friendship with Schelling at the time of these lectures, this philosophy had a deep correspondence with his own method of researching and pointed the way beyond merely intellectual thinking. While Kant viewed the mind as structured according to its own categories of understanding (causality, necessity, unity etc), Schelling (like Fichte) saw the power of mind as in no sense limited to judgements dictated by its own categories. The basis of Schelling's *Naturphilosophie* is that 'the system of nature is at one and the same time the system of our mind' meaning that we can actually *think* the living processes of nature. Reason, for Schelling, is not merely an ordering power of mind but is of the nature of life itself, because mind itself is organic. 'As long as I myself am identical with nature, I understand what living nature is as well as I understand myself.'[48] Out of such views Schelling prophesized a future poetics of nature uniting the aesthetic and the scientific.[49]

Schelling's philosophy was highly influential on the thinking of Wilhelm von Humboldt who was given the task of developing the founding principles of the Berlin University and who was its director of education from 1809 to 1810. This university is widely held to be the first modern university and universities all over the world today still look to the 'Humboldtian university' as a model and reference point. Humboldt had considered the ideas of Schelling as well as a detailed proposal put forward by Fichte but in the end it was the suggestions of Schleiermacher that the founding principles resembled most closely.[50] However, the influences on Humboldt had not just been philosophical because he had had, in effect, a training in the practice of Goethean science which had transformed his own approach as a scientist. During the years 1794 and 1795 Humboldt and his brother, the geographer Alexander, were in Jena working on a daily basis with Goethe on the dissection and observation of anatomical specimens. Wilhelm von Humboldt's understanding of Goethean science was profound and through the agency of his ideas this way of science came close to the cultural threshold where it could be admitted into the university curriculum.

Goethe had already been preoccupied with questions of animal and human form for almost two decades before working with the Humboldt brothers. Rudolf Steiner writes that in 1776 Goethe's 'sculptural spirit' was at work, trying to understand the morphology of animal skulls. The approach he was taking was to see the outer sculptural form of the skeleton as a means of knowing the inner being of the animal, just as in human physiognomy, the facial features are seen as an expression of the inner life or soul.[51] In 1781, in relation to his studies of human anatomy, Goethe wrote that he was treating 'the bones as a text to which all life and everything human can be appended'.[52] As the Humboldts engaged with Goethe in this work and realized its significance they urged the poet to put his ideas into writing, which he did, although not in any comprehensive form. These studies had the greatest impact on Wilhelm von Humboldt who went on to apply Goethe's approach to anthropology and later to history, fields not entered into by Goethe at all.

It was Goethe's phenomenological approach which inspired in Wilhelm von Humboldt the idea that by comparing different examples of animal skulls for instance, we can learn to read archetypal forms, hidden to the external senses but active as the 'formative ideas' which give rise to these visible phenomena. What Humboldt discovered in this time spent with Goethe was that *a new form of thinking* was required in order to understand the forms of living things, even animal corpses or skeletons which are the product of living growth processes. The German scholar Chad Wellmon writes about 'the efforts of Humboldt's lab partner in Jena, Goethe, to imagine a new scientific method and, more precisely, a new form of observation that might adequately account for the dynamism of nature'.[53] This new form of observation was the method Goethe had been working with for years, which he called 'beholding thinking' (*anschauende Urteilskraft*)—thinking *in* things through cognitive feeling. In other words, by working with Goethe, Humboldt was receiving a schooling in what Goethe called 'exact sensory imagination'. Wellmon writes:

> Such a mode of observation acknowledges that the very act of observation is always already bound up with the natural objects. And this recognition entails an imperative to make our concepts as dynamic as nature itself. Nature as a unity is given or rather gives itself to us: therefore, our own thinking must become adequate to nature.[54]

Wellmon goes on to discuss how Wilhelm von Humboldt's anthropological research developed on precisely the same basis, seeking a new scientific method involving this same mode of observation. Humboldt does this in his 'Plan for a Comparative Anthropology' (1797) in which he relates his method to Goethe's work with plants and his notion of the archetypal plant; Humboldt seeks to observe humanity as if it were a 'giant plant'. Humboldt transposes Goethe's imaginative morphological method with plants onto the human form, so that any particular human racial formation becomes a kind of 'text' of form through which the archetypal human form can be 'read'. In other words, like Goethe, Humboldt is seeking to grasp the *formative* processes in nature which lead to diverse human races.

In 1799, ten years before the founding of the University of Berlin, Humboldt wrote his *Aesthetic Essays I: On Hermann and Dorothea* in which he concisely describes the role of the artistic imagination in cognition, clarifying the approach which is fundamental to Goethe's way of science—that the artistic element is not merely fused to or adjunctive to the scientific, but that science realizes itself as a true *life* science through the artistic imagination. The cultivation of the imagination in this way, Humboldt writes, develops 'completely other organs than those that usually lead our lives'.[55] Later, in the early 1800s, he sought to apply the same dynamic, formative approach in his study of language, focussing on the Basque language, the study of which he announced in 1812.[56] It should be mentioned here that Humboldt's ideas on this subject have a resonance with those of Friedrich Schleiermacher whose views, as noted above, most closely guided Humboldt in developing the foundations for the new university. Schleiermacher also saw the artistic way of interacting with natural form as a model for the science of anthropology.[57]

Around a decade after the founding of the university, in 1821, Humboldt set forth the same Goethean approach, this time in relation to the study of history.[58] Humboldt could perceive in this field the same problems which Goethe saw in the biological sciences of his time: that the real living nature of a phenomenon is only half understood or distorted by a mode of observation which stops at superficial appearances. He is seeking to 'make visible the truly activating forces' within the chaos and seeming arbitrariness of external events; in other words he is proposing a kind of physiognomy of human history.[59] Humboldt's statements are entirely in accord with the way Goethe carried out his phenomenological studies of organic forms: '[The historian] has to seek the truth of an event in a way similar to the artist's seeking the truth of form'; historians should be endeavouring to grasp 'creative forces in world history'.[60] Humboldt writes further:

> It may seem questionable to have the field of the historian touch that of the poet at even one point. However, their activities are undeniably related. For if the historian . . . can only reveal the truth of an event by presentation, by filling in and connecting the disjointed fragments of direct observation, he can do so, like the poet, only through his imagination. The crucial difference, which removes all potential dangers, lies in the fact that the historian subordinates his imagination to experience and the investigation of reality. In this subordination, the imagination does not act as pure fantasy and is, therefore, more properly called the intuitive faculty or connective ability. . .[61]

Here we have an extraordinarily clearly delineated application of Goethean science to cultural phenomena. In the entire disquisition a highly developed understanding of Goethean methodology shines out; it is obviously written by someone who understands not merely the philosophy but the practice of this science, who has actually made the effort to find the unity of the scientific and the artistic in his own way of working. Humboldt evidently knows the difficulties and challenges of carrying out this kind of research authentically: 'This freedom and subtlety of approach must become so much a part of [the historian's] nature that he will bring them to bear on the investigation of every event.'[62]

To repeat: in the founding of what became the model for the modern university—the Humboldt University of Berlin—a Goethean artistic form of science came close to entering into mainstream cultural life and was certainly in the background of the founding impulse. The fact that after its founding the history of the modern university took the pathway already noted—eventually towards economic rationalism and a neo-liberal form of organization—does not alter the fact that with this first modern university the ground for a Goethean form of science was prepared. An externalizing impulse was at work, a working outward of forces in civilization seeking the integration and healing of the rift between the artistic and scientific, of the division between humanity and nature. We can imagine that, even with the immediate relationship of Goethe's way of science to the ideas of German 'nature philosophy' which were abounding at the time, it was still too fresh, too radical, to become the working methodology of a new university. In fact, the form of the new Humboldtian university was radical enough and broke with tradition in profound ways, in particular in connection to the role of research within the university context and the central place of philosophy.

There is an intensified interest in the Humboldtian university in the English-speaking world today, as part of the soul-searching concerning the fate of the modern university and the envisioning of future possibilities. In many quarters a lamentation is being expressed over the destruction in recent decades—through the rise of economic globalization, neo-liberalism and post-modern scepticism—of the Humboldtian idea of the university as a core institution of democracy and human spiritual development.[63] Related to this we are seeing efforts to restore the key values and methods of the Humboldtian university, such as the link between teaching and research at all educational levels.[64] There is no question that the value and meaning of culture, as espoused and promoted in the original Humboldtian university, are largely lost to the contemporary world. 'Culture' for the Humboldtian university ideal means a particular form of cultivation (*Bildung*) leading to the highest aim which has to do with the realization of the human individuality: self-knowledge as the basis of world knowledge. What is meant here is not merely a 'philosophical position' but the realization of the human being as a unified whole—the intellectual, the moral and the physical.[65] Humboldt wrote:

> In order for an individual to extend and individuate his character, he must first know himself, in the fullest sense of the word. And because of his intimate contact with all of his environment, not only know himself but also his fellow citizens, his situation, his era.[66]

This was the research ideal as originally conceived; it was not intended as an exclusive right of academics over and above their tasks as teachers but as an integral part of the whole educational experience of the students. The genuinely inspired search for knowledge is that which gives meaning to the whole higher educational experience—not merely taking the form of a limited 'research proposal' which can easily become artificial through having to fit externally-imposed funding requirements and the obligations of academic tenure. In Humboldt's words, all aspects of higher learning should be individually motivated and 'creatively sought in the depths of the human spirit'.[67] Something of the essence of these ideas and values is what must have inspired the thousands of American scholars who travelled to Germany in the middle years of the nineteenth century and who transplanted what they experienced into the American university system.[68] And it is the withdrawal of the contemporary university from these high cultural ideals which has led the academic commentator Bill Readings to claim that the university now lies in ruins.

For, indeed, aims such as the enlightenment and ennoblement of the student are now considered to be entirely old-fashioned and redundant. The distinguishing feature of the modern university is not cultivation but rather the concept of *excellence* which doesn't originate in the cultural-spiritual sphere but has a techno-bureaucratic basis.[69] The university begins to define its mission and success in terms of 'performance indicators' in accounting terms borrowed from the economic sphere (efficiency, excellence, pertinence etc), this being one expression of the progressive corporatization of the university in the contemporary world. Bill Readings writes: 'The social responsibility of the University, its accountability to society, is solely a matter of services rendered for a fee.'[70] In the absence of the unified body of teachers and scholars in the traditional sense, the body corporate assumes that role—the corporation—and the university is subsumed into the economic sphere of society and is evaluated through econometrics.

Rapidly increasing numbers of young people worldwide are now seeking to undertake degree courses at university. The modern multi-versity campuses swell with fee-paying students and the diversity of courses on offer is impressive—because, of course, these are not physical ruins. The ruins of the ideals of truth, knowledge, self-realization, goodness, beauty, cultivation, are nevertheless present. Readings reconciles himself to the situation as follows:

> We have to recognize that the University is a *ruined institution*, while thinking what it means to dwell in those ruins without recourse to romantic nostalgia or fervent modernization. The trope of ruins has a long history in intellectual life.[71]

The key question becomes: how to dwell creatively in the ruins in order to conceive and bring into being the university of the future—this is the positive face of postmodern subversion. Ruins have a rawness—something is stripped away. One experiences a nakedness in standing before the remains of life without the fervent busyness and day-to-dayness which hides the real character of things. There is even something awe-inspiring about ruins when we reflect on the immensity of the human creativity which has gone on over centuries, in building up the forms which presently lie in decay. Contemplating ruins can be a source of inspiration and a stimulus to our sense of responsibility—in the spectacle of ruination we may consider how an opportunity presents itself for something new to come into being.

Let us imagine that we are wandering around the ruins of the university—not the physical ruins but ruins in the sense of the subverted idea of the university, which require more of an imaginative eye to be perceived —a little like Goethe as he wandered on a moonlit night around the ruins of the Greco-Roman civilization during his sojourn in Rome in 1786. In his first *Roman Elegy* he wrote:

> Tell me you stones, O speak, you towering palaces!
> Streets, say a word! Spirit of this place, are you dumb?
> All things are alive in your sacred walls
> Eternal Rome, it's only for me all is still.[72]

In the quiet and evocative ruins he senses something new which he has the feeling will involve him intimately. 'But soon it will happen, and all will be one vast temple/Love's temple, receiving its new initiate.' It's possible, too, to go with our eye of the imagination in the quietness of contemplation amid the ruins of the university and feel something like a reaching out of new life, like a new love, the possibility of something which was only ever longed for in times gone past. For at this moment the need for intellectual cleverness is wiped away and the words of the heart are what sound out.

At this moment, these are the questions which may spring up: what do we most care about at this juncture in the history of humankind? What is most deserving of our solicitude? On our contemplative walk everything is seen in a new and almost uncanny light; the spiritual treasures of the past gain a new effulgence. It becomes clear that what can come into existence can only grow from the condition of things as we find them in the here and now. What has gone before—the medieval community of scholars, the old *studium generale*, the Humboldtian vision of cultivation, the reconciliation of science and art—none of these things are actually negated but are alive in the ruined walls and stones. All is ready to be brought to a new level. And as we contemplate these 'open secrets' of the past on our metaphorical walk through the ruins of the university, we can as well give deep consideration to the metamorphosis of the university and what it would mean to teach Goethe's science of living form within a university of the future.

Beyond 'The Doctor'

Goethe's science of living form does not lend itself to a doctoral form of education and the Goethean scientist, strictly speaking, is not a doctor. Now, this may appear as a rather extreme position given that the doctoral form of education, originating in the medieval universities, has stood the test of time. It could be argued that it is quibbling when the important thing is not the title but what an individual achieves through their teaching and research. But names need to be accurate and we tend to take the office of 'doctor' for granted when considering what the modern university stands for. In fact 'doctor' has a particular connotation relating to a definite educational ideal; it points to the fundamental intellectuality of university education—today just as in the medieval period.

The word 'doctor' comes from Latin verb *docére*, meaning 'to teach' and in the Middle Ages the first doctorate *(licentia docendi)* meant the graduate was licensed to teach, having completed studies in the 'higher faculties' of medicine, law and theology. If we look at what was involved in attaining this degree we can come to an idea of the form of learning which stood behind the doctoral ideal of education. Students were expected to understand and retain what they had learnt and this, in the case of law for example, was the study and memorization of the treatises on Roman Civil Law of Justinian I and Canon Law as codified by Gratian in his *Decretum*.[73] The doctor, as 'the one who knows', was a person who had attained a mastery in this respect. This medieval tertiary education was through and through intellectual in nature with a particular emphasis on the logical: the practice of logic was at its heart and was what shaped its every procedure. For example, in theology, proofs of the existence of God were a central preoccupation. Aristotle's treatises on logic set the standard of rational analysis and, in particular, his work *Organon* provided a detailed treatment of syllogism (the fundaments of logical structure) and analysis of proof.[74] In the founding of the universities in Europe it was thus the cultivation of the *thinking* function of the human being which was the central motivating factor.

To understand the metamorphosis of educational ideals in Western civilization it is necessary to penetrate to the underlying formative, structuring forces at work. For this we must reach further back in time, to the most archaic periods of human life when there was not the division between education and the rest of cultural-spiritual life as we know it today. With the earliest peoples education was not about knowledge in

the sense of intellectual understanding but more what we would desig-
nate by 'know-how', emphasising the *doing* aspect. Tribal initiation—for
example, with the Australian aboriginal people—concerned how to go
about life in accordance with methods worked out in time immemorial:
where to find food, how to remove poisons from certain plants in order
to make them edible and so on. Centrally, the ritual life had a strong
active aspect in dance, the sacred dances in which, in a bodily way, the
initiate connected with the gods. Archaic education was not a matter of
philosophically pondering the nature of the divine but rather of enacting,
through ritual life, a specific relationship with particular gods. In Ancient
Indian culture the bodily doing aspect was refined to a high degree in
the practices of yoga—abstinence, breath control and the carrying out of
body positions called *asanas*. In general terms we could say that it was
the doing or *will* aspect which was emphasized at this stage of human
development.

When we look to the ancient Greek educational practice, as presented
in Plato's *Republic* in particular, we still find the will aspect to be fundamental. This text is read today as a work of philosophy, and rightly
so in a certain respect. However, we misunderstand the *Republic* if we
assume it is merely an intellectual production. It actually sprang from and
gave expression to an educational process which was essentially bodily in
nature. As Rudolf Steiner points out, in the ancient Greek world there was
a strong link to the East—that is, to the yogic bodily form of education; for
indeed, ancient Greek culture stood as the bridge between East and the
West and was actually an offshoot of oriental civilization. In the *Republic*,
the basis of educational practice is music and gymnastics—that is, harmo-
nized movement with the aim of achieving the total harmonization of the
human being.[75] Steiner observes:

> [The Greek] was only concerned to develop the human body in such a way
> that, as a result of the harmony of its parts and its modes of activity, the body
> itself should blossom into a manifestation of divine beauty. The Greek con-
> fidently expected of the body just what we expect of the plant: that it will of
> itself unfold into blossom under the influence of sunlight and warmth, if the
> root has received the proper kind of treatment.[76]

The condition striven for through Greek education was not some kind of
'knowledgeable person' in the modern sense but a quality of beauty, proportion and balance which we see brought to expression in the sculptural
works of Classical Greek culture.

Steiner notes how easy it is, with our intellectualism, to look down on
the Greek form of education and its ideal of the Gymnast, in particular
how it was practised through athletic exercises (the *palestra*) and the arts
of dance and gesticulation (the *orchestric*) which were conducted accord-

ing to definite musical-rhythmic principles. This, Steiner says, is a result of misunderstanding how the Greeks saw the fruits of cultivated bodily expression. At about the age of twenty, when the student's education passed over to the study of geometry and mathematics, it was considered that these subjects grew out of the bodily education: 'It was known that geometry is inspired in the human being by right movements of the arms.'[77] Likewise with other elements of Greek tertiary education, all parts of the seven-stage *trivium* and *quadrivium* were built upon the inner sense of proportion and harmony inculcated through the bodily arts of movement and dance.

In the ensuing developments of the medieval period in European civilization, the ideal of education gradually shifted its focus, to that of the Rhetorician. This again is something hardly understood today because of our tendency to look down on earlier ideals of education from the 'heights' of our modern intellectualism. It is not easy for us to understand how and why the ancients conceived the aims of a rhetorical education, beginning in Roman times, which now made the speech art of rhetoric primary and the bodily arts secondary. Rhetoric, one part of the preparatory *trivium* for the ancient Greeks, became all-important in Roman times and was cultivated in the monastic schools and in the universities, through the writings of Cicero's *De Inventione* in particular. Through rhetorical speech an appeal was made to the *feeling* life of the student, not primarily to the logical mind. This is not to say that the teacher-as-rhetorician in this period was not expected to have an extensive knowledge derived from the various fields of learning; indeed, it was Cicero who brought into existence the notion that the rhetorician (as the 'ideal orator') could not use 'empty words' but, rather, a speaking which must draw upon a wide-ranging humanistic understanding of the world.[78] But the appeal of this teaching was made primarily to the students' feeling life in order to convince them of the truth of whatever was being taught. It was the Church orator who became the crowning glory of this form of teaching during this period.

The aim of education through rhetoric still carried the impulse of doing or willing, now not so much bodily but in the sense of a will-to-convince. The convincing of others through speech is a manner of doing but at the same time the word 'takes on the colour of thought' as Steiner puts it.[79] Thus the ideal of the teacher as Rhetorician is a middle position between the earlier ideal related most fundamentally to the will (the Gymnast), and the ideal which would come later and which is related most fundamentally to the thinking (the Doctor). Even in the earliest medieval universities such as the University of Paris, it was the Rhetorician who reigned supreme, the one who was most skilful in argument and who was a master of dialectic. But as the universities developed through the middle period of the Middle Ages another ideal emerged, that of the Doctor, which gave the highest place to the intellectual development of the human being. Numerous influences led to this, and we have already seen that one of them was the new awakening of the power of logic through the rediscovery of the ancient texts of Aristotle.

The rise of the ideal of the Doctor or Professor marks the beginning of the modern period. The bodily aspect ceased to have any value, but just as little the ability to convince as something good in and for itself. What now is valued takes place within the human being invisibly, in the aspect of the inner life we call the intellect; *what is known* becomes the primary focus of education. The first expression of the doctoral ideal in the Middle Ages was a focus on what is called encyclopaedism which values the breadth of a person's book learning. The ideal of the Doctor celebrates the one who has a vast knowledge in their particular field and the student's mark of success was assessed along the same lines—by memorization and examination. The idea of a student sitting for months and working with a pile of books in order to be tested by someone as to how much they knew—this would have seemed complete nonsense to the ancient Greek working within the ideal of the Gymnast and equally to the Rhetorician. Most contemporary forms of educational theory and practice continue to stress just this form of learning and examination; this medieval ideal is still largely the norm in our time, no matter how far transformed other aspects of education are in relation to traditional practice.

If we trace the metamorphosis of the doctoral ideal into our own time we come to the advent of the computer and the so-called Information Age. If we follow this metamorphosis of educational ideals inwardly it becomes clear that the computer is the most powerful contemporary manifestation of the doctoral ideal. Computers are a focalized expression of what is most essential to this ideal: the computer has an encyclopaedic memory and works in a purely logical fashion to make information readily available through the Internet and other forms of interactive technology (in a transformed sense, the computer 'professes' its store of knowledge). If we study the computer as a phenomenon we will gain a considerable understanding of what the 'the doctor' actually signifies in its essential form. Indeed, it is not going too far to say that the computer *is* the Doctor in a new embodiment of itself—which means a pure intellect, nothing whatsoever to do with a bodily basis (the will) or a life of feelings.

As stated above, Goethe's science of living form does not lend itself to a doctoral form of education and the Goethean scientist is strictly speaking not a doctor. It can now be seen more clearly why this is the case; it is so because this science is not fundamentally an intellectual form of endeavour. Whether the student is considered to be a passive receptacle into which knowledge is poured or whether the student is seen to learn best through interactive experience and 'problem based' methods makes no difference to the fundamental intellectuality of the doctoral educational style. By contrast, Goethean science calls for *a thinking of the whole human being* meaning that the thinking, feeling and willing as a unity are involved in the cognitive process, not merely the intellectual mind. It cultivates forms of thinking which have no place or even meaning within doctoral education—namely, the capacity for *cognitive feeling* and *cognitive will*—both of which will be explored in the following chapters.

The new ideal of education is what Steiner has called the ideal of the 'whole' or 'universal human'; this, he says, has emerged in the modern period into ever greater prominence.[80] We have already seen how, in different epochs in the development of Western civilization, three different aspects of the human being became emancipated from the original undifferentiated nature of archaic spiritual life—firstly the bodily or will aspect, then the feeling life, and most recently the thinking aspect or intellect. The ideal of the 'universal human' draws all three into a unity on a new level, not through some kind of 'integral theory' of education, but through actually *realizing* a human thinking in which the power of feeling and willing is made active. The teaching of Goethean science is central to the developed ideal of the whole human being because it provides a pathway toward a genuinely living, unified form of thinking.

The new educational ideal of the 'universal human being' develops *out of* the doctoral ideal because it heightens and transforms the thinking aspect. This is the key point: *through this heightened thinking capacity, feeling and willing themselves become organs of thought*—thus we can speak of 'cognitive feeling' and 'cognitive will'. This consciously achieved metamorphosis of thinking represents a great challenge for those involved in the task of transforming the universities. Just as the doctoral mind tends to look down on the earlier educational ideals of the Gymnast and the Rhetorician with a sense of superiority, so this doctoral mind has a natural difficulty in accepting the idea that there is an educational ideal which goes beyond it and will eventually supplant it. The challenge is profound and far-reaching in its implications—the metamorphosis of the doctoral ideal is nothing other than the metamorphosis of the university.

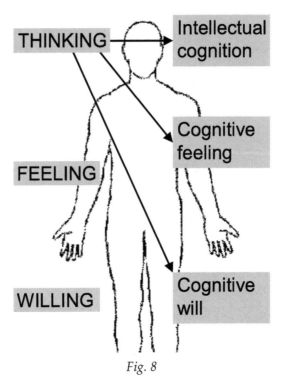

Fig. 8

PART II
FOUNDATIONS OF A TERTIARY EDUCATION IN GOETHEAN SCIENCE

'. . . a methodology of university study must be rooted in actual and true knowledge of the living unity of all the sciences, and ... without such knowledge any guidance can only be lifeless, spiritless, one-sided, limited'.

Friedrich Schelling[81]

'As human beings, we must not allow ourselves to be tyrannized by academic knowledge. In our efforts to emancipate cultural activity, we are combatting the abstract character of academia as such and placing human beings first . . . Humanizing academic activity is our goal. We must work toward bringing the human being to the fore in so-called objective scholarship which must be grounded in life and in human beings. Those of us who engage in it must not become dry and shrivelled. On the contrary, by "combatting abstract existence", as I call it, we become useful contributors to the very necessary process of counteracting the barbarization of Western civilization.'

Rudolf Steiner[82]

'It sounds almost paradoxical to say that the hearts of university students should be addressed in everything that is brought to them. It sounds like a paradox, but it actually could be so!'

Rudolf Steiner[83]

Chapter 4
The University Curriculum and Community

Orientation Studies

In relation to the university of the future which is inspired by the ideal of the universal human being, we must conceive a new course of orientation studies which specifically cultivates the key elements of a genuine life science—that is, Goethean science. What is offered in this section are indications concerning how such key principles and methods may be taught; the intention is not to set forth a fixed and final picture of the form of a foundational course.

The risk is that what is meant here by orientation studies will be interpreted as a form of undergraduate core curriculum or course of breadth subjects in the way these are normally meant in contemporary universities. Whether it is compulsory core or elective breadth subjects, the aim is essentially the same—namely, to explore and communicate a certain intellectual content. This by no means encompasses what these orientation studies set out to achieve. If the content of the next three chapters—polarity, intensification and metamorphosis, imagination and delicate empiricism, type and archetype—were merely made the subject of an intellectual form of teaching, then no new point of departure would have been found and nothing advanced. An orientation study in Goethean science seeks through its intellectual content (which is here called 'intellectual cognition') to orient the mind towards a living, imaginative form of thinking, towards the threshold to the sphere of living form—that is its sole intention.

In a certain sense universities were at a related point at their inception, in the early medieval period, when the doctoral ideal inspired the task of educating the mind towards a logical intellectual form of thinking. Intellectualism was certainly not the general characteristic of the pious religious consciousness of the time. Already, in Chapter 3, we have noted how logic was not just a subject of study in the first universities but pervaded all other subjects as a method. The kind of rational, logical thinking which is so familiar to us today, inculcated through school education where young people are taught to think through ideas and issues for themselves, was for the medieval mind only a possibility, a potential of extraordinary significance. The point we stand at today with regard to higher education is related to the task of the first universities only in this respect: a new educational ideal demands a dedicated, focussed and obligatory study to cultivate a form of thinking which for the culture of our time is something new and exists largely as a potentiality.

Today the classical doctoral core study—the original liberal arts curriculum—is no more. It was gradually transformed even during the Renaissance period, modern content was introduced and it was finally subverted by the postmodern onslaught which determined that no fixed, compulsory course of general study was acceptable. The fact that the curriculum forming the very core of doctoral education *could* be subverted is significant as a symptom of the evolution of human consciousness. It points to the fact that the doctoral ideal in tertiary education has in its most essential aspect been consummated. At the threshold of the sphere of living form the meaning of a compulsory course of study gathers to itself an entirely new impetus. Thinking must consciously, *out of its own will,* bring about a self-transformation, a metamorphosis to a further possibility of itself. The educational ideal of the whole or universal human being can only come alive as a conscious, willed creation of the individual human spirit. In the first universities the curriculum was an act of collective dedication to the wisdom of bygone thinkers and their great books.[84] In the university of the future the curriculum has meaning only as an individual act of responsibility to the living being of the Earth.

Tertiary students in the future will need to understand all these matters; understanding the aim and the form of their education will have to be an essential part of that education. In a way it is an affront to the intelligence of young people entering university to suggest that nothing in relation to their non-professional studies can be obligatory and directed. For many, perhaps most, what they are seeking in institutions of higher education as they set out on the path of life as adult human individuals is *guidance* on their quest of learning. Young people will have no problem with a compulsory orientation study, they will never feel it as something imposed on them, so long as they are helped to understand the reason for its existence. They will need to be carefully taught about the evolution of educational ideals; they will need to discover the difference between inorganic (mechanical) and a living form of thinking. And especially, they must learn why their orientation studies in Goethean life science have significance for the whole of their education, including their professional studies.

So let us consider more closely what is involved in these orientation studies, why they involve so much more than an intellectual content. The students are taught to see how the intellectual content passes, as it were, into the whole human being, how the intellectual aspect is taken up into the feeling and will. In the initial teaching and learning practice, this is a definite sequential process in which the thinking transforms the feeling into the faculty of cognitive feeling and transforms the will into the faculty of cognitive will. To demonstrate this we can follow an 'elementary' example of metamorphosis—a theme of the next chapter—which is nevertheless profound in its implications.

The example we will be dealing with is the metamorphosis of plant form. In the first stage of intellectual cognition the notion of metamorphic process in nature is presented. This is a theme which is wide-reaching both scientifically and philosophically; philosophically it relates to the meaning of evolution, to the processes of formation and transformation throughout the natural and human worlds. Students will naturally have the opportunity to discuss and explore this theme but they need to do something more as well—they need to become conscious of the quality of detachment which is the character of intellectual cognition. The intellect feels itself to be separate, removed from the things it is trying to understand. In intellectual thinking the logical mind moves in a world of concepts of an abstract but clear and definite nature in which we experience the freedom of objectivity. Self-knowledge is awakened; we become aware of the way in which we wake up to the world through our intellects but nevertheless remain detached from it.

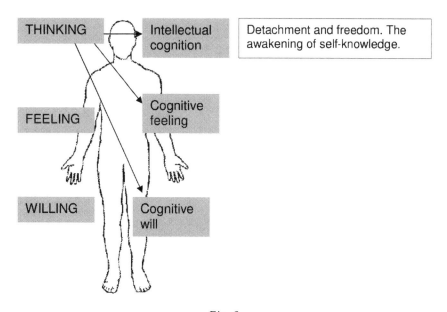

Fig. 9

Self-knowledge now shows the path to a further stage in the quest for knowledge; we seek to unite with that from which we are separated through our intellectual/logical form of thinking. Now begins the form of thinking we are calling 'cognitive feeling'. We may consider a basic and well-established exercise in Goethean science—the perception of metamorphosis of leaf shape. Let us contemplate the following two leaves from the same plant—the first a lower leaf, the second a leaf not far below the flower (there are a number of intermediate leaf forms).

In a Goethean plant morphological study we 'pour ourselves' into the exact form of each leaf (just as we do any other part of the plant we

Fig. 10

are studying). This is no vague, facile requirement; it is actually very exacting and difficult to carry out. Only in our feeling capacity are we able to 'enter into' and assume the precise form of the leaf. A student may well ask: Which aspect of our mind is active here? It is certainly not the intellect—it is *an exact form of feeling*.

Cognitive feeling is beyond sympathy or antipathy; the act of assuming the precise form of the plant—the leaf, the stem, the flower or any other organ—is a pure deed of feeling because the plant itself is pure, without the slightest trace of sympathy or antipathy. When feeling is consciously heightened to become cognitive feeling there is a

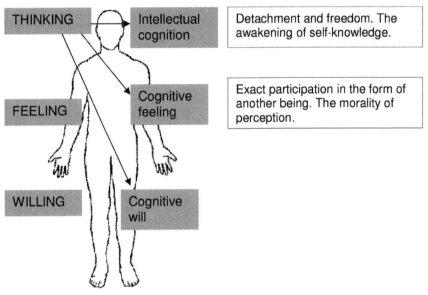

Fig. 11

purifying process involved and this we may call *morality of perception.* Students need to become aware that the character of cognitive feeling is essentially moral. When they actually experience this character they will then appreciate how the feeling life—under normal life circumstances moving unconsciously and constantly between sympathy and antipathy—can unite itself morally and cognitively with the form of another being.

Self-knowledge guides us further along the path of knowledge for we have entered consciously into the form but not yet reached the formative *activity* of that form. We move to a consideration of not just the form of the leaves but the formative process which has formed them and which unites them. We have mentally participated with exact feeling in the form of each leaf and now we can move from one to the next in order to inwardly experience the transformation. We cognitively 'live in' this metamorphosis, we enact it in a precise, definite way in the exact imagination. Knowing now becomes a form of creative activity; we are mentally willing or 'doing' the metamorphosis of the plant.

Thus we speak of 'cognitive will' at the stage in which we are consciously engaged in the creative, formative process of organic form; it is *an exact form of will.* In this way the student comes to experience and understand a form of creativity which is selfless or non-egotistical—for indeed, the form of a plant is not the expression of a creative self or ego. This stage of thinking can be called a *morality of creation.* We have already seen, in Chapter 2, that in order to understand the formative 'idea' in nature, the *typus,* we must creatively and comprehendingly 'do' it. This stage of an orientation study prepares the mind for further tasks of cognitive will which may be part of a student's professional studies and, indeed, their professional work altogether.

Now it is possible to see why these orientation studies are something entirely different from core or breadth studies. Just through this 'elementary' example of leaf metamorphosis it becomes evident how far beyond intellectual content this study has taken us. If these stages

Fig. 12

have been genuinely experienced then the fundamental aim of the Goethean orientation studies will be understood: to cultivate a way of knowing the world which is at the same time the moral development of the human being. In this it fulfils, in a new way for the future of human civilization, what has always been the essential task of the university—namely, to enlighten, to provide circumstances which make possible the getting of wisdom. Throughout the history of education the moral-spiritual

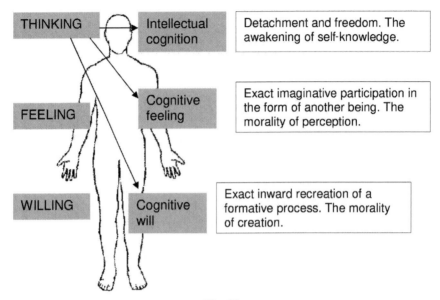

THINKING — Intellectual cognition — Detachment and freedom. The awakening of self-knowledge.

FEELING — Cognitive feeling — Exact imaginative participation in the form of another being. The morality of perception.

WILLING — Cognitive will — Exact inward recreation of a formative process. The morality of creation.

Fig. 13

development of the human being is the ideal which has continued to shine out even when everything else around it has fallen away into decadence.[85]

The basic form of these orientation studies set forth in the following chapters corresponds to the threefold character of the sentient life of the human being—the thinking, feeling and will. All three elements are nothing separate, any more than the head is separate from the heart and hand in terms of the physical body; they are three different aspects of a unitary being which is the human soul. Here is the structure in a more developed explication:

- *intellectual cognition:* relates to the more conceptual aspect of cognition and thus embraces the study of philosophy in a broad sense. Here the issues and questions connected to cognitive feeling and cognitive will are considered in a theoretical way: the meaning of living thinking, the nature of mechanism and organism, metamorphosis etc, moral and social issues in the modern world, the meaning of self-knowledge, the form of the curriculum and ideals of learning, the philosophy of science and art. At the centre of this is the philosophical adumbration of the fundamental principles of the science of living form.

- *cognitive feeling:* the practice of the science of living form (Goethean science) stands as a central mediating role through the way it unites the scientific and the artistic. The scientific element of Goethean science connects it most closely with intellectual cognition; the artistic element of Goethean science connects it with cognitive will. The fundamental aspects of life science are taught as practice through delicate empiricism and exact sensory imagination: polarity, metamorphosis, intensification, type, archetype and archetypal phenomenon.

■ *cognitive will:* the practice of arts such as modelling, painting and poetry. All the concepts and elements of Goethean science (metamorphosis, polarity, intensification, archetype etc.) are cultivated purely artistically, through modelling, colour work, musical expression—forms of artistic practice which lend themselves to such expression. The aim is not 'self-expression' nor the development of art as such but art only in the service of knowledge. Art is of the same order as creative process in nature; the artistic sensibility raised to a higher possibility of itself becomes cognitive will.

Although the elements of Goethean science are discussed separately here, in practice they are not necessarily taught as discrete 'subjects' but could be introduced into the form and content of orientation studies which are of a university's own choosing. For example, a university may elect as the make-up of its orientation studies: psychology, ecology, drama and fine arts (painting, modelling, sculpture), architecture, philosophy and politics. Into each of these subjects could be woven the concepts and practices of gesture, polarity, intensification, metamorphosis, type, archetype and archetypal phenomenon.

These orientation studies would continue throughout all the years of a tertiary education; they are not the same thing as introductory or prerequisite courses which cover general and relatively elementary content in preparation for more specialized professional trainings.

A Phenomenology of The Outer and Inner Worlds

Goethe once commented that he had long been suspicious of 'the great and important-sounding task: "Know thyself!"' He considered this to be something confusing in its impossibility, diverting attention from the outer world onto 'some false, inner speculation'.[86] This injunction was said to have been inscribed above the portal of the Temple of Apollo at Delphi, an ancient mystery centre, and was the basis of the Platonic philosophy of education. Plato, as Rudolf Steiner notes, still had one eye looking to the transcendental spiritual ways of self-realization of the East even while the other looked toward the birth of Europe and intellectual, scientific forms of knowing.[87] Know who you are in your essential nature, in your higher selfhood—such was the substance of this ancient exhortation. In several places and in the light of the wisdom of Socrates, Plato points out the absurdity of trying to know things about the world when one has no understanding of oneself.[88]

It might be asked: should not the same words be inscribed above the entrance to the contemporary university, for is it not true that all human beings are on a quest for self-understanding? Is not everyone seeking to know for the sake of what—or who—they are living? This is surely so— yet the modern university is a secular institution without the theological orientation it had in the medieval period. Young people leaving school are seeking to find their way in a difficult world; they want to know what to do with their lives. But that doesn't mean they consciously and specifi- cally desire to 'know themselves', to take part in a schooling in mediative inwardness as something distinct from learning about the world and how to act constructively within it, which is their actual curriculum of study.[89] Goethe declared: 'The human being knows himself only insofar as he knows the world . . '.[90] Know thyself! could perhaps be written above the portal of the modern university but only if it refers to the kind of tertiary education in which knowledge about the external world yields knowledge of the self.

To be sure it sounds paradoxical to speak of knowledge of the world and self-knowledge as one and the same thing. In terms of the doctoral ideal of education with its emphasis on the intellectual or purely objective way of knowing, this notion verges upon the absurd. However, the pathway to self-knowledge through world knowledge becomes apparent in relation to the educational ideal which supports the whole human being becom- ing active in cognition—the thinking, feeling and the will. Conventional science considers things purely objectively which means that 'self' and 'phenomenon' must be carefully isolated from each other—and we have already considered in earlier chapters how this represents a necessary step on the path of knowledge. But we have seen that an authentic life science takes us into deeper dimensions of the phenomenon through a conscious, participatory way of knowing. Beyond the threshold to the sphere of living form the faculty of *cognitive feeling* awakens; feeling becomes a power of cognition to the extent that it identifies *exactly* with the dynamic, gestural nature of the phenomenon. Further, *cognitive will* discovers that the cre- ative forces at work in the world are the same forces at work in the human being. In the teaching and practice of Goethean phenomenology at the ter- tiary level, a division between external knowledge and self-knowledge is not possible. The phenomenology of the outer and inner worlds represents a unity; both 'lookings' are part of the same study and are shaped con- jointly within the curriculum.

The human inner life or sentience is constituted by the thinking, feeling and the will. Thinking is the polarity of willing within the wholeness of sentient life. If we consider the phenomenon of thinking in its pure form we see that it has the character of clarity and definiteness. Our primary experience is that our thoughts and concepts are not personal in character, not bound up with our own selves—for example, the concept of a plant or a mathematical theory. Concepts are associated with definite mental pictures and this is why thoughts have a universal character and is why we call them 'objective'. Only by virtue of this objectivity, this impersonality, do we speak of thoughts as having the character of *truth*. The things and processes of the world are detached, lawful and self-sufficient; they appear to us as having their own self nature and not as an extension of our own self. We do not produce the external world—it just presents itself to us. Likewise, we do not produce meaning, we do not create truth—these just become apparent to us. We sense our thought life in relation to the head—the brain and associated sense organs—for the human head itself, in relation to the body as a whole, has a detachment and freedom granted by our upright posture. Unlike animals, the human head 'rides' on the body in a relatively unmoving fashion and is served by the arms which are also made free by this posture.

By contrast, what we mean by will is something individual, something bound up with our own selves. The act of willing to do something does not have the luminous character of thinking for while willing might *follow* from reflection, it is immediate and direct and has to do with our activity. Much of what we call will is unconscious: the coordinated movements of our limbs and the activity of metabolism, things which we are driven to do through instinctive urges and strivings. The German philosopher Arthur Schopenhauer wrote: ' [Will] appears in every blindly acting force of nature, and also in the deliberate conduct of man . . '.[91] In one way or another will has to do with the impulse to action which transforms and which creates; with the human will we always speak of *experience*, not reflection, objectivity and truth. The impulses of will do not 'become apparent' to us like thoughts or mental images—we ourselves carry them out. Further, the aspect of the world which is shaping itself from within through formative, growing, metamorphosing processes—these are the forces we experience as will. They are not the mechanical forces which come to light through logical thinking; these are creative forces which can only be experienced creatively, as a language of gesture enacted by the will. The same gestural language which is experienced within things exists within ourselves, as our own will forces. We sense the will in connection with our limbs and with the dynamism and 'fire' of the metabolic and reproductive organs.

The aspect of human sentience which lies 'between' the life of will and thinking is feeling; feeling has characteristics of both willing and thinking. In the first place, feeling has something of the personal and dynamic character of will; the word 'e-motion' itself carries the idea of action. Feelings surging up from the unconscious will we experience as antipathies; these feelings draw us away from the surrounding world and isolate us. On the other hand, feelings can 'go out' and enter into other things in a more conscious way, engaging sympathetically with the creative processes at work in them.[92] Feelings (like will processes) are not just mental images but activities. In our feeling life we do not *reflect* on phenomena like we do in thinking—we enter into things, just as water and air can surround and permeate things. The phenomena of the world are not impenetrable presences; even the rocks and minerals have an open, yielding character and the formative forces at work within them can be experienced. But there is also something objective about feeling; we *carry out* deeds of the will but we *have* feelings just as we *have* thoughts; feelings become apparent to us. Nevertheless we do not speak of feelings as 'true' as we do with thoughts; feelings are experiences. We experience our life of feeling in the central human being, in relation to the principal organ of breathing and the heart—both dynamic, rhythmical organs. Just as the breath connects human physiology to both the inner and the external worlds, so feelings mediate between the objectivity of things and inward urges of the will.

The educational ideal of the whole human being will never be realized in a merely theoretical way; it must be a practice.[93] In an orientation course of studies based on Goethean science, a phenomenology of the inner and outer worlds is developed at every point and interstice of the educational journey. It is like Goethe's experience with his colleagues during his adventure in learning in the artistic colony in Rome which he likened to 'a room full of mirrors where, whether you like it or not, you keep seeing yourself and others over and over again'.[94] When one studies a mineral, a plant or animal to the point of cognitive feeling and cognitive will, when one enters with cognitive imagination into the nature of the human being and human society in its multifarious forms and expressions, one is constantly coming upon a mirror to the self. Goethe founded the science of the living world in which knowledge of the world is knowledge of the self, and this can and should be the discovery of every student in the university of the future. It is upon this basis that all forms of professional study can be shaped.

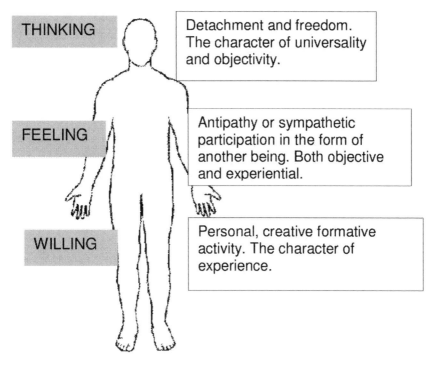

THINKING — Detachment and freedom. The character of universality and objectivity.

FEELING — Antipathy or sympathetic participation in the form of another being. Both objective and experiential.

WILLING — Personal, creative formative activity. The character of experience.

Fig. 14

Goethean Science and Professional Education

The fields of knowledge in their present form, the university faculties, have been developed for centuries as intellectual structures in terms of the doctoral ideal, as creations *of* the doctoral ideal. The inspiring ideal of the whole human being will lead to each field of knowledge undergoing a profound metamorphosis—but this will be a lengthy process. Some areas are already advanced in this direction; the study of medicine, for one, has been fructified by Goethean phenomenological approaches out of which has developed a methodology of symptom pattern recognition in the clinical setting.[95] Goethean approaches have been brought to bear on numerous aspects of the humanities and biological sciences.[96] In relation to economics a Goethean research study is presented in Chapter 8, set forth as an exploration of why the social sciences, including economics, constitute true life sciences. How these Goethean scientific studies apply to the physical sciences like physics, engineering and geology or to subjects of a purely spiritual-cultural nature like law and politics, are questions which will also need to be addressed.[97] To do so will require a great breadth of vision in relation the whole structuring of future tertiary curriculums.

We may take as two examples the specialist scientific studies of geology and optics. Geological structures and processes are in themselves lifeless but in terms of the entire 'body' of the Earth they are aspects of a living whole; that is how Goethe himself approached the study of geology.[98] Likewise with the study of optics, a branch of physics: seen from a limited logical and mathematical perspective, colours appear as the products of lifeless processes, as mere quantities—that is, wavelengths. Within the broad vision which imaginative understanding permits, colours appear as manifestations of the polarity of light and darkness, as qualities or characteristic 'gestures' which relate to each other meaningfully. As with the other areas, it is the approach of whole-seeing which characterizes the Goethean understanding of colour phenomena.

In general, we can say that what is understood within the doctoral mode of education to be strictly and narrowly mathematical will need to be re-envisioned from within the sphere of the living whole. This is so in the case of pure mathematics and the highly mathematical disciplines of physics and engineering. Even inasmuch as these subjects are developed mathematically they can be grounded in a curriculum of goetheanistic orientation studies. Rudolf Steiner's observations help to clarify the rationale behind such an approach:

> Mathematics deals not with things, but with those aspects of things that lend themselves to measurement. And it must acknowledge that this is only *one side* of reality and that there are many others over which it has no control. Mathematical judgements do not fully encompass real objects; they are valid only within the intellectual realm of abstractions that we ourselves have conceptually isolated from the full reality as *one* of its aspects. Mathematics abstracts the magnitude and quantity of things . . . It would be a great error, however, to believe that mathematical judgements can encompass nature in its wholeness. Nature is not merely quantity but also quality, and mathematics limits itself to the quantities.[99]

This, we may say, constitutes a starting point which opens vistas on how the new orientation studies can develop in relation to particular specializations. If indeed mathematics is but one aspect of reality, involving logical powers which are but one dimension of the human soul, then it *must* be presented to students in that light; this is the imperative and teaching responsibility born of the educational ideal of the whole human being. In this way the different specializations—whether they be in the physical, social or biological sciences or in the humanities—are integrated within the organic totality of the future university curriculum. Nothing of the vast knowledge already built up in these specialist fields of learning is negated. To repeat what was stated in Chapter 1: the power of a truly living thinking, a thinking in which the whole human being is made active, can take hold of this knowledge as a revitalizing, synthesising force. The living being of nature which 'dies' in abstract theoretical knowing is brought to a new life.

In the future there will be methods and activities connected with professional studies which today would strike us as very unusual; they will bear little resemblance to the current courses of study which are built from intellectual content and have a familiar style of delivery involving lectures and seminars, text readings and written assignments. The birth of radically new forms and ways of working are going to require a considerable period of gestation. Still, it is important to even now form living pictures of what a new form of tertiary education may look like because it is only through such creative imagining that this education can sooner or later come into being. For students of economics for example, the orientation studies would begin with phenomenological research, including an exploration of human and natural forms like plants and the human skeleton, colours and water phenomena, in which polarity, metamorphosis and intensification are worked with in depth. These understandings would then be progressively related to social and economic phenomena. The Goethean study of economic phenomena documented in Chapter 8 is situated at approximately the grey spot in Fig. 15, mid-way between Goethean natural scientific studies and the field of professional economic science proper. Teaching at this mid-point serves as a transitional function for the students, to gradually allow for living ways of thinking and perceiving to fructify within the field of economic science as a whole.

It is now possible to present diagrammatically and as a general picture the form of the proposed orientation studies in relation to the specialist faculties (see Fig. 15). If we consider the three dimensions of the cognitive mind as a whole then we can see how the core curriculum radiates out towards the different specialized fields of learning, integrating them. The circle is symbolic of the university as an organic whole within which the human being is imaged. Thus we see that the form of the 'organic' university curriculum is an image of the whole human being.

The Living Community of Higher Learning

It would be nonsensical to consider placing a curriculum for the cultivation of living thinking into a social and physical context which does not support that curriculum, ideally in every respect. Just as the parts of an organic form such as an animal are interrelated through each being a reflection of the whole animal, so in the living form of the 'organic university' every element— from teaching and assessment modes to the architecture which clothes this teaching—can be imbued with the impulse of life.

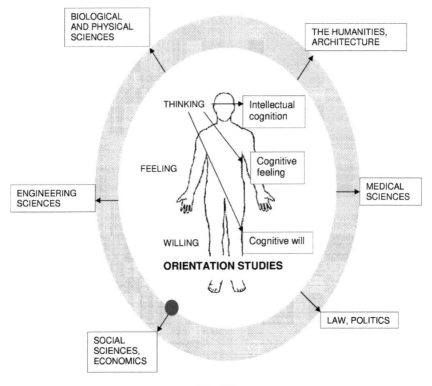

Fig. 15

In terms of the architectural shaping of the new university much work in the direction of a 'living', dynamic form of architecture could be referred to.[100] The aspect which is of immediate concern here is the creation of dynamic, vibrant communities of learning through which the ideal of the whole human being may continue to shape itself and radiate its significance into the broader cultural life of society.

When one develops an eye for it, it becomes possible to read the doctor principle written in every aspect of the learning environments of contemporary universities; it perhaps stands out most clearly in the relations between students and teachers and in the teaching methodologies. Academic elitism or the 'ivory tower' phenomenon, the detachment of teachers from students, the tendency to value private research over teaching—all these things have been critiqued over recent decades yet without specifically connecting them to the doctoral ideal. In fact, these are all manifestations of a one-sided concentration on the thinking aspect of the human being as the locus of all truth—that is to say, of intellectualism. Detachment, objectivity, cold-hard factuality, prosaic theorizing—all these are the expressions of the ideal of intellectual learning and each is a signature of the doctoral conception of higher education.

The idea that tertiary education is fundamentally about the acquisition of knowledge through inquiry and study, and peripherally (or not at all) about self-knowledge and community, is very much bound up with the doctoral ideal. Intellectual learning in universities was disposed from the beginning towards disputation and this gave rise to the modern lecture and seminar system with its methodologies of instruction and debate. On-line learning and the electronic transmission of lectures has now become a significant part of a transformed tertiary education but that doesn't mean that the traditional doctoral *form* of lecturing has changed at all. Lectures are normally regarded as the way students know what will be examined in any course of study; a certain content of knowledge is delivered, often highly prosaically, and notes are taken. Yet, in the age of the Internet, this content is highly accessible in ways other than listening to a lecture—either live or recorded—which means that the value of maintaining this form of instruction is very much in question. This situation also throws into relief the issue of what university teaching staff are going to teach if attendance at physical lectures is quite evidently unnecessary. But doubt about the value of the doctoral form of lecture is nothing new; in Rudolf Steiner's review of 1898 entitled *University Education and the demands of the Present Time,* he acknowledges the uselessness of lectures which take the form of lengthy elaborations of objective facts which can be found anyway in what he calls 'compendiums' or 'manuals' (in other words, library resources). Rather, there should take place relatively brief 'orientation lectures' which present general perspectives—'great, essential vantage-points according to [the lecturer's] personal understanding'.[101] What Steiner was highlighting over a century ago is something which is now blindingly and incontestably apparent; what took the form of 'compendiums' in his time is today what we call 'on-line access'. He goes on to discuss how the seminar work should be expanded and, right from the beginning of university studies, should focus on developing research skills.

The doctoral form of lecture is redundant but that doesn't mean the lecture itself is outmoded; in the living community of higher learning, inspired by the ideal of the whole human being and dedicated to the cultivation of a living form of thinking, lecturing methodology itself needs to undergo a metamorphosis. By looking back at what was exceptional and groundbreaking in the teaching style of the philosopher and academic Johann Fichte, Steiner gives a glimpse at what we could look forward to in the organic university of the future:

> [Fichte's] purpose was to awaken spiritual activity and spiritual being. From the souls of his hearers, as they hung upon his words, he sought to call forth a self-renewing spiritual activity. He did not merely communicate ideas.[102]

Through presenting 'great, essential vantage-points' the lecturer can open out to the students, to the greatest extent possible, the kind of awakening, living insights which the students can then take with them as inspirations into their own research in more focussed ways. But this also brings up the question as to what sort of qualifications a person needs to become a university teacher and the associated issue of tertiary teaching training—beyond merely acquiring a doctorate and having writings published.

The nature of debate associated with the doctoral form of seminar puts great onus on each individual being able to present their views strongly and to reply to other points in a decisive and convincing way. This mode of learning, by emphasizing the power of the intellectual mind, tends the learning experience toward the competitive and away from the kind of reflection which gives rise to self-knowledge. Let us consider the way of learning involved in the cultivation of cognitive feeling. As we have explored in Chapter 2, this has to do with developing the capacity to enter *into* another being and to think *within* the form of that being—whether it be a plant, animal, another human being, a phenomenon of the economic realm or a work of art. It is of the nature of feeling in its cognitive capacity to be selfless as water is 'selfless' and flexible, to assume the form of another being with heightened consciousness and sensitivity. Such learning is for this reason disposed towards a capacity for *listening*; feeling participates in the form of another being through a kind of 'listening thinking'. The conscious and intentional form of receptivity we are calling cognitive feeling balances and transforms the intellectual mind. Goethe himself stressed the need to cultivate conscious receptivity in scientific research in order to avoid the dangers of preconception and rigidity.[103]

The practice of Goethean science engenders a particular kind of social practice in a seminar situation; Christina Root writes of so-called 'Goethean conversation':

> The ideal of conversation becomes a model for Goethe of a kind of multifold language that can overcome some of the difficulties inherent in applying a particular method to phenomena . . . The Goethean method calls for continuous self-examination and self-transformation in just the way a good friendship does. Developing a multifold language is part of the process by which we train ourselves to see from 'the perspective of objects', and learn to imagine ourselves empathetically into the position of our partner in conversation.[104]

When such conversation happens between people in an educational situation—for example a group who are engaged together in the study of a phenomenon from the natural or human worlds—we have the potential for a truly intimate community of learning. And in such a learning community the psychological space is created in which self-knowledge and knowledge of another are understood to be one and the same thing. Paradoxically enough, in the moments of greatest selflessness the understanding of self is greatest; when the restless waters of the egocentric, grasping 'I' are stilled we can see ourselves best. Such intimate 'conversations' are possible with plants and animals, with other humans, with all the world's phenomena, and the building of such relationships constitutes the way of Goethe's science of living form. Thereby it becomes possible to gain some inkling of what the German poet and philosopher Novalis was alluding to when he spoke of a higher community of 'men, beasts, plants, stones and stars'.[105]

This consideration of an enlivened social context of learning is not a reference to the general social life in and around campus but to the *learning experience itself*, the way education can happen. A consciously created, engaged social form of education arises when human individuals, each sovereign over themselves, learn together in a way which is attentive to and respectful of each other's individuality while conscious of the great questions which hover before humanity as a whole. The tone of such community is set by the connection of teacher and student because here is where the quality of mutually supporting human relationship can shine out most strongly. The mega-proportions of modern tertiary institutions and the demise of programmes of studies which are in common for all students from all faculties make such relationships difficult or impossible; students can undertake a whole degree over several years without ever experiencing what so stimulated Goethe in Rome—a vital, productive, ongoing community of learning. We can only look towards the kinds of small, intimate, self-governing tertiary institutions, each unique in emphasis and consciously promoting truly human values and relationships, which Paul Goodman hoped for decades ago in his *The Community of Scholars*.[106]

Chapter 5
Gesture, Polarity, Intensification & Metamorphosis

Introduction

What follows in the next three chapters are the essentials of a Goethean science teaching methodology. Gesture, polarity, intensification and meta-morphosis, exact sensory imagination and delicate empiricism, the law of compensation of parts, type and archetype, are key 'letters' in the language of living form which Goethean science is concerned to decipher. They are fundamental principles for understanding the organism, each being understandable only in terms of the others. As suggested in the previous chapter, they need not be taught as separate subjects but, rather, could be woven into the content of an orientation studies programme. The artistic examples provided here for teaching Goethean science are only possibili-ties; many others could be worked with from the realms of mineral, plant, animal and human phenomena.

The organization of this and the following two chapters is related to the sequential stages of the orientation studies, as outlined in Chapter 4. These stages are intellectual cognition, cognitive feeling and cognitive will; together these constitute the thinking of the whole human being.

Intellectual Cognition (Philosophy)

By intellectual cognition is meant what would be called in another setting 'theoretical considerations' or 'philosophical perspectives'. This provides a background to Goethean science (in this chapter gesture, polarity, inten-sification and metamorphosis) and an introduction to what can only be understood through phenomenological praxis.

In one place Goethe writes how our ancestors admired the economy of nature, how life develops from a few basic principles which are diver-sified by intensification 'into the most infinite and varied forms'.[107] Right away Goethe's thoughts strike a note which is unfamiliar to our modern ears, ears which are so used to hearing about Darwin's theory of evolution as the incontrovertible truth. We have become habituated to the idea that everything in nature is produced through the material and largely chance processes of natural selection. The idea that there are creative 'basic prin-ciples' at work in the development of natural form is most certainly at odds with Darwinist theory, even anathema to it.

Two of the basic principles Goethe is referring to are polarity and intensification or enhancement (*Steigerung*); he speaks of these as 'the driving forces in all Nature' which, he says, can be perceived.[108] The word 'perception' (*die Anschauung*) is important here because we need to be careful lest we assume Goethe was merely speaking of what is sense-perceptible. Goethe was without doubt an astute observer of the physical forms which make up the natural world but he also meant perception in a 'higher' sense. He was referring to the capacity of imaginative thinking to 'see' into the living, creative processes of nature.

Polarities are not the same as opposites—or at least, they are opposites of a particular kind. The theory of opposites belongs to ways of thinking coming from the proto-philosophical world and set down by Aristotle; before him opposites had figured in the ideas of pre-Socratic thinkers including Parmenides, Heraclitus and Empedocles, all of whom Aristotle acknowledged. Tall and short, hot and cold, wet and dry—Aristotle commented about such opposites: 'They must be of the same kind, though *of* that kind, as different as possible'.[109] Opposites can be blended to produce a gamut of in-between qualities or shades; for example, hot and cold mingle to produce degrees of warmth (or degrees of cold). Polarities, by contrast, do not merely blend but have a dynamic, creative relationship. Goethe writes that the union of poles occurs 'in a higher sense' where intensification has taken place, whereby the intensified poles 'will produce a third thing, something new, higher, unexpected'.[110] For example—and this was a key realization in Goethe's experimental work—light and dark do not merely mingle and neutralize—they unite to produce something which is fundamentally different from the two polarities—colour. An example of polarity in the human world is male and female, the union of two (in sexual reproduction) producing a 'third thing', something new, not just a blending or neutralization of the two.

Goethe notes that it was through Kant's writings that he was drawn to the idea of attraction and repulsion in the natural scientific outlook; he realized that 'neither can be divorced from the other in the concept of matter' and this insight led him to the recognition of polarity 'as a basic feature of all creation'.[111] Kant's writings on this theme had a profound influence on other thinkers of the time as well. The philosopher Friedrich Schelling, who for a period had a close connection with Goethe, expounded a view of polarity in terms of the absolute principles of self and nature. Kant had reduced all matter to the polar forces of attraction and repulsion and Schelling extended such thinking to a complete picture of the whole of nature working organically, whereby the forces are at one point in balance, at another productive of new physical developments. It was Schelling's view that the concept of the organic was necessary for understanding the non-living, in which attraction and repulsion 'structure a complete system of formative powers'.[112] Through his association with Goethe he arrived at the insight that the polarity of attraction and repulsion in the total coming-into-being of nature relates to the organic forces of expansion and contraction—for example, as seen in the growth of a plant.[113]

Polarity is at work everywhere in nature but at certain points it shows itself most clearly—it comes to the surface, as it were—and one such focal point is the magnet. It was for this reason that the magnet was of considerable interest to Goethe and his colleagues.[114] Schelling connected his observation of the physical phenomenon of the magnet with what he called the 'intellectual intuition' of polarity as a formative principle at work in universal nature. Writes Schelling: 'Matter as a whole is to be viewed as an infinite magnet.'[115] Kant had already come to the idea of the *copula* (Latin: 'that which binds') as the reciprocal relation of two polarities; for Schelling the copula became fundamental to his thinking on the dynamic equilibrium of nature.[116] In essence this is the idea that every polar relation is a threefold—the two opposing polar principles and the link between them, this being the third principle. With the magnet we have the forces of attraction and repulsion relating to each of the poles and we have the force-field relating the poles. The magnet becomes a model for the structure of polarity as such and this, for Schelling, becomes significant in relation to his theory of identity and his ontological world-picture. He explores the idea of the copula as the mediating principle in the polarities of freedom and necessity, self and nature, subject and object, being and becoming.[117]

When we grasp the threefoldness of polar relationship (two poles and the relating principle or copula), we are on the way to understanding intensification and metamorphosis in relation to nature's formative processes. In the history of philosophy it is only recently coming to light to what extent Schelling was indebted to Goethe in relation to the development of these ideas.[118] Schelling came to view the whole of living nature as a dynamic equilibrium which means something capable of transformation and metamorphosis, driven by a series of 'potentials' (including magnetism, electricity and chemistry) which all have polar opposition within them.[119] Nature, he realized, is productivity and product; productivity has development and progression or what we may call intensification. Because of the polar oppositions inherent in nature there can never be stasis, for the tendency in one direction or another is always brought back into equilibrium through that which unites the poles (the copula).

It was Goethe, through his studies of natural phenomena, who arrived at the point of being able to perceive the relationship of polarity and intensification in natural formation. In the physical form of the plant his 'eye of the imagination' discerned the creative activity of polarity—in the growth of a plant in particular. Between the poles of Sun and Earth the plant begins its life as a seed form belonging to the Earth, and realizes itself in the floral form which, qualitatively, belongs to the Sun. Earth, in this perception, intensifies towards the Sun and the changing leaf forms reveal this metamorphosis. What is inchoate in the seed is brought to expression most fully in the flower where fertilization takes place. We can speak of the 'spiritual striving' of nature on the level of the plant, inasmuch as all of life is an expression of the polarity of spirit (idea) and matter.

Thus the plant becomes the image for intensification as it works throughout nature. Writes Rudolf Steiner:

> What Goethe calls enhancement consists of the bringing forth of the spiritual out of the material by creative nature. That nature is engaged 'in an ever-striving ascent' means that it seeks to create forms which, in ascending order, increasingly represent the ideas of things even in outer manifestation . . . The creative spirit of nature comes to the surface of things here; that which, in coarsely material phenomena, can only be grasped by thinking, that which can only be seen with spiritual eyes, becomes, in enhancement, visible to the physical eye.[120]

Nature is not something static; it is constantly developing, metamorphosing and producing. The philosophies of Kant and Schelling clarify much in relation to what Goethe meant when he said that polarity and metamorphosis are 'the driving forces in all Nature,' productive of all form. Such philosophical considerations prepare us for the kind of nature study which seeks to cultivate our powers of observation. Philosophy stimulates and clarifies our thinking in relation to the broad concepts of being and becoming in nature and the question of how knowing takes place. However, it is only through phenomenological work that the faculty of cognitive imagination awakens, the 'living thinking' vital to the development of a true science of living form. It is to this that we will now turn.

Cognitive Feeling (Goethean Science)

a) Gesture

As teachers we are guides on the pathway from intellectual cognition to cognitive feeling; this pathway leads us to the phenomena themselves. We may, for example, work with the example of the common dandelion.

The phenomenon that we have in front of us is a complex entity of many parts with a unique form. To take in the form we need more than our eyes; the plant has a characteristic scent, taste and texture. All of this we observe in the most careful way possible, respectful of its uniqueness; we observe stems with a certain size and texture, leaves with a particular shape and size, flowers with a unique form, colour and scent. We recognize that all these parts comprise the entity we call *Taraxacum officinale* and which we classify as belonging to the Asteraceae family within the plant kingdom.

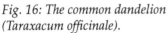

Fig. 16: The common dandelion   Fig. 17
(Taraxacum officinale).

It is valid and necessary to regard the dandelion plant in this detailed, factual way; but it must be acknowledged that we would describe any non-living material object such as a crystal in the same way. How do we bring these perceived facts alive in our minds? This is another way of asking: how do we see this organism as a living whole? We do this, not by changing the facts or theorizing about the facts, but by allowing our *thinking* to come alive. In this way we are drawn towards the threshold to the realm of living form.

As we move our feeling-thinking through the different facts and observations we have made we begin to unite them, using the comparative method to see each as a reflection of the unity which is the whole organism. The more we do this in an open, receptive way, the more possible will it be for musical insight into the dandelion to awaken. We see now that there are two significant points of radiation in this plant—the leaves all branch off from a point just above the root, and the flower parts all radiate from a central point—this can been seen even more clearly when the flower becomes its seed form. We also experience radiance in the gold-yellow of the flower—it would be quite different if it was blue or purple. This is the beginning of insight into the characteristic dynamic quality of the whole plant and a first step in our cultivation of a living thinking. The physical eye sees the radiation as a perceptual fact, part of a normal scientific description of this plant. The 'eye' of cognitive feeling sees the same fact as a living *gesture,* which, being a dominant or signature feature of this plant, points the way to a deeper understanding of the whole organism. To understand polarity, intensification and metamorphosis in living form we need to learn how to read the language of living gesture.

Fig. 18: Imperfect and perfect cadences

What is a gesture? A definition is not good enough; it needs to be cognitively *experienced*. For this we can turn to actual music, for any musical composition is formed in its entirety of different kinds of gestures; music is a language of gesture. Cadences are very specific and common gestures within classical music; the imperfect cadence (I-V) expresses a resolution which is 'suspended'—that is, it is not conclusive; the perfect cadence (V-I) brings the time-form of the music to a definite and complete conclusion.

This language of gesture doesn't just belong to music. It doesn't matter whether it is visual form (as with the plant) or auditory form (as with music), the 'eye' of cognitive feeling perceives form as dynamic and meaningful. A gesture communicates a specific meaning although that meaning may be subtle and only possible to be understood in relation to the total gestural language of an organic form. Thus the gesture of radiation of the dandelion is only meaningful in terms of the dynamic character of the whole plant; and the cadence in terms of the whole piece.

Let us turn to the art of sculpture where gesture is not expressed through time as with music, but where gestures are 'held' in three dimensional spatial forms. Sculpture nevertheless expresses dynamic (and in this sense musical) meanings.

Fig. 19

This sculpture (Fig . 19) is composed of two polar gestures in close relationship. One is free and embracing, the other is self-enclosed and embraced by the greater. They tend towards each other in such a way that a dynamic of relationship is created in the space between; this third relating or mediating element is an actual sculptural experience although physically invisible—it is visible to the 'eye' of cognitive feeling. We can call this the uniting gesture. The work as a wholeness is thus a threefold gestural relationship.

b) Polarity and Intensification

Art is the province of cognitive feeling; nothing in the realm of art may be understood without it. Through our experience of the sculpture we can now turn back to nature and contemplate the phenomenon of the fluid substances air and water. Here polarity can be expressed in a particular way—as stasis and movement—which under certain conditions come into dynamic relationship. We can picture a situation where a still body of water has a wind moving over its surface and see that two poles are uniting at the boundary between them. Here the poles do not merely blend and neutralize like opposites but, rather, give rise to a 'third thing,' the uniting element or copula, which is the phenomenon of *rhythm*. The whole of this phenomenon is inwardly pictured and experienced in cognitive feeling as a threefold gestural relationship.

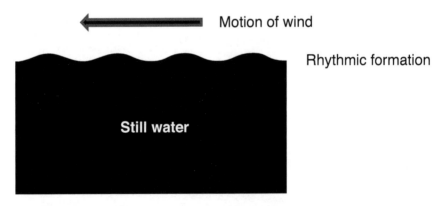

Motion of wind

Rhythmic formation

Still water

Fig. 20

In the world of organisms polarity works productively leading to complex, individualized formations. The plant stands between the earthly and the cosmic, between Earth and Sun—therein lies its polar nature of which the physical plant before us is an image; it is also imaged in the plant's reproductive process.[121] In the flowering plant this polarity comes to overt expression (compared to, say, mosses and ferns). One pole, comprising the seed and the root, belongs to the realm of Earth and expresses many qualities which unite it with the Earth's mineral body—the hard particle-like form of the seed, the root's linear formation penetrating the soil, the bitterness of the root. The opposing pole is the flower, with its soft, coloured, perfumed, chalice-like form exuding the sweetness of the nectar, altogether expressing its belonging to the realm of the Sun and the sunlight.

The plant *grows*; that is to say, the productivity which takes place through the union of the poles of Earth and Sun is of an extremely developed nature compared to, say, the productivity of the poles of the magnet. Still, the same archetypal principle of polar threefoldness can be perceived in the plant world through the practice of cognitive imagination. The growth of a plant is progressive in time—it is not merely a matter of enlargement. The creative intensification which takes place between the poles of Earth and Sun produces a metamorphosis of form in which creative power—the invisible 'in between' of the copula in other realms of nature—becomes something *physically visible*.

1. POLE (SUN — THE FLOWER)

2. COPULA (RHYTHMIC FORMATION OF VEGETATION WHICH UNITES EARTH AND SUN)

3. OPPOSITE POLE (EARTH — SEED AND ROOT)

Fig. 21: The flowering plant.

Let us look to music again to aid our understanding of organic growth. Intensification becomes visible in the change of form through time—that which we discover in plants is clearly evident to cognitive feeling in musical experience. In the following section of a melody we do not perceive merely a rhythmic repetition of the same note from the beginning to the end of the section; rather, something develops, something is created through a process of metamorphosis. One tone gives rise to the next tone progressively with a directionality which gives meaning to the melody. Growth, as we experience it musically, is not a matter of enlargement but of an unfolding in time. In music we hear growth.

Fig. 22: Excerpt from Für Elise *by Beethoven.*

c) Metamorphosis

Turning back to the plant world we can consider flowering plants in which the 'signature' of productivity (that is, growth) is manifest in a particular way. These are the herbs which reveal a strong metamorphosis of leaf forms (in this case, hedge mustard). If we have understood how the gestural language of music reveals growth as a directionality in time, then it will be easy to arrive at a musical perception of metamorphosis in leaf form. To do this we enter into the changing forms with cognitive feeling and inwardly enact the metamorphic process which takes place between Earth and Sun (as with the two leaves presented in Chapter 4). The metamorphosis moves from the rounded leaf forms closest to the Earth pole, through progressively more divided and differentiated forms, to the pointed forms held close to the central stem near the Sun pole. Intensification drives metamorphosis.

In the geological realm metamorphosis means that one rock type turns into another. This is not the case with metamorphosis in plants; one leaf in the sequence does not physically turn into the next. Each leaf in the sequence becomes and stays what it is. What then is the *movement* of metamorphosis? This is a movement which cannot be physically perceived yet it can be inwardly grasped in an exact way. Cognitive feeling assumes the configuration of a leaf and then flows to the next; cognitive will experiences the continuity of form as a creative or formative activity.

Every one of the millions of species of plants is an image of the threefold polarity of Earth and Sun but, as we have seen, some plants heighten and make visible this threefoldness to a much greater degree than others. In some cases this threefoldness is expressed in a onesided or highly specialized way—and it is often this onesideness in plant form which gives rise to the nutritive or medicinal properties of those plants.[122] Sometimes the Earth pole is especially enhanced and has been further exaggerated by human breeding—for example the carrot and beet, where qualities of sweetness and colour which normally belong to the Sun (flowering) pole of the plants are 'pushed down' into the realm of the root. In some plants the Sun pole is enhanced—for example, the cauliflower where the blossom is drawn down into the vegetative realm. In the case of the mint family of plants

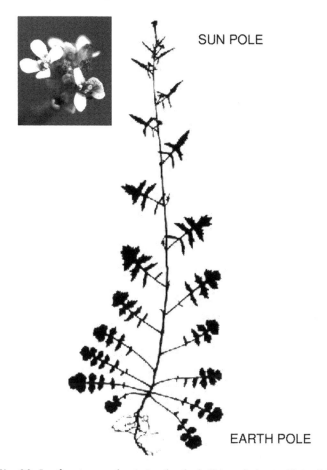

SUN POLE

EARTH POLE

Fig. 23: Leaf metamorphosis in charlock (Sisymbrium officinale).

(*Labiatae*), every one of its 3,000 species has medicinal properties.[123] Here the leaf forms are highly specialized and emphasized in the sense that, while the flowers themselves are relatively inconspicuous (diminished), the Sun pole 'flower process' which is expressive of warmth and light has been pushed into the leaves where it takes the form of 'fiery' aromatic, volatile oils; the leaves become thereby highly medicinally enhanced.[124]

In every sphere of nature we find polarity; likewise in the human realm—in sexuality and physiology for example—polarity is everywhere active (in the Case Study documented in Chapter 7 polarity in human physiology is explored). Helped by artistic work students can come to experience how polarity is generative of form and how all transformations of form take place within the 'creative tension' of the poles. Polarity has an archetypal nature and the particular examples which we can work with in an educational setting are what Goethe called 'archetypal phenomena'—that is, manifestations which clearly reveal the nature of the archetypal driving forces of nature. However, the aim of an orientation course in Goethean science is not a specialized study of any particular phenomena; rather, the particular cases or archetypal phenomena can open the way for the student toward self-understanding—for polarity, metamorphosis and intensification stand behind all human existence. Here we may refer again to the statement of Goethe's quoted in the previous chapter: 'Human beings know themselves only in so far as they know the world ...'

Cognitive Will (Artistic Practice)

Metamorphosis exercise:

The aim of this exercise is to create a metamorphic series, modelled in clay, starting with one geometric or organic shape and ending with another. Only the starting and ending shapes are decided on at the beginning. There could be six or seven steps involved. The eight or nine sculptures could be more or less the same size or an expansion or contraction could be involved. Each sculpture could begin as a sphere formed between the palms. This work needs to be carried out slowly and deliberately, entering with cognitive feeling into each form in relation to the other elements of the series so that the practice becomes something very different from a mechanical process or logical puzzle. One should, right from the outset, sense the potential sequence as a wholeness so that each form at each step is created as arising out of that whole rather than just through a visual juxtaposition—this is what makes it an artistic practice.

Sculptural polarity:

This clay modelling activity involves fashioning two polar forms within the one sculpture and then relating them in different ways. It starts with perhaps half a normal block of clay; this piece could be firstly shaped into a sphere. The main thing is that this work starts with a strong sense that the clay is a 'one' or a whole and this continues even when it is divided. Out of this initial 'oneness' two polar forms gradually emerge—this polarity could involve high or low, pointed or rounded, concave and convex etc. As the two poles emerge they are not merely formed separately but are always related artistically so that the oneness of the clay becomes a three-fold relationship—a 'one which is a three'.

Plant metamorphosis imagination:

The metamorphosis of leaf shape in herbs follows the same essential pattern as that of hedge mustard: rounded germinal leaf forms which stretch out and differentiate higher up the plant before becoming pointed and contracting upon the main stem near the flower. In this drawing or painting exercise a flower can be created from the imagination and placed at the top of the page with a stem connecting it to the earth (variations could involve colour, size, tubular or opened shape, number of petals, number of stamens etc). Then the leaf metamorphosis can be created from the imagination, but with cognitive feeling always sensing the changing leaf forms in relation to the already-created flower so that the intensification toward that particular flower is meaningfully expressed in the leaf forms.

Threefold painting or drawing:

The important point in this exercise is the initial and continuing sense of the sheet or canvas as 'the whole' out of which individual elements emerge. This is an abstract colour work in which a threefold relationship is developed. A number of different colours are used but each colour should be carefully selected through cognitive feeling to create a very conscious gestural language for the whole work (shape and texture can also be involved). Two separate areas should develop polar qualities through the use of colour and these should be related through an in-between or mediating area which both separates and connects the poles.

Case Study: Polarity, Intensification and Metamorphosis in The Realm of Colour

The Goethean observation of colour can be a central exercise in the orientation studies we are considering here. In many ways this kind of colour work serves as a foundation for considerations of plant and animal form, and of the constitution of the human being. Colours have a purity of presence, an immediacy to the observing eye, unburdened by form, texture, scent and so on. Even the simplest plant or animal is a much more complicated phenomenon. It is not difficult to assist students to arrive at the point where they can say they have *experienced* a colour whereas, for example, with the animal studies developed in the next two chapters, reaching the point of cognitive experience is not so easy. We cultivate cognitive feeling in a most powerful way by entering into colours and their relationships.

The Goethean approach to the understanding of polarity, metamorphosis and intensification in relation to colour is fundamentally an education in *perception*. The kind of perception we are helping our students with has already been briefly explored in Chapter 2 with a look at the colours yellow and blue. There it was discussed that, in order to experience the colours in terms of themselves, to read the 'language' of colour, we must learn to be able to participate in the phenomenality of the colours by feeling our way with cognitive imagination into the inner activity or gesture of a colour. It might well be asked: how can a colour have a gesture? Whatever else may be said intellectually or theoretically by way of answer to this question, the essential thing is that it is *perceived* to be the case. The eye which perceives the specific gesture of the different colours is not the physical eye but the 'eye of the imagination'.

More precisely, there are different stages in colour perception. Firstly, we have a visual, sensory impression of a colour—we simply look at it and recognize it just as we look at and recognize objects in the world around us when they are not part of any conscious experimental work. Next, if we consciously 'enter' and permeate the colour with our feeling (or, perhaps better put, when we allow the colour to permeate our feeling); we may now speak of cognitive feeling. If this is deepened it reaches into the will, meaning that with our creative sensibility we participate in the 'doing' of the colour in an exact way (the colour gesture or inner activity). Now we may speak of cognitive will. Passing through all these stages the whole human being is made active in cognition.

Goethean colour studies lead to the understanding that colour arises as a 'third thing' through the interaction of the polarity of light and darkness. Now, of course, this is not what conventional science tells us; for this science, darkness is merely the absence of light. Discussion can certainly be had about whether colour is 'contained in the light' and separated out by diffraction through the prism (as the physicist Newton believed and as it is still normally considered today) or through a union of light and darkness as Goethe considered to be the case.[125] It must always be borne in mind that Goethean science proper is *phenomenological*; to cognitive feeling, darkness is no mere absence of something else but a phenomenon in its own right, a gestural presence equal and opposite to that of light. Intellectual thinking may insist that only light has being and reality but for cognitive feeling the polarity of light and darkness is incontrovertible.

Although light may surround us and illuminate objects, light itself has in no sense a perceptible existence. As the philosopher Hegel wrote: '. . . just as little is seen in pure light as in pure darkness'.[126] Hegel demonstrates philosophically that darkness cannot be merely the absence of light, that both light and darkness must have real existence in that they determine each other mutually; it is the same with the poles of a magnet. White and black tend toward the nature of colour in that they have sense-perceptible existence; they are images of light and darkness. When we contemplate white and black with feeling-thinking we are led to the inward experience of light and darkness as polarity.

How one comes to perceive the production of colour from the interaction of light and darkness, using the prism, is something documented well elsewhere.[127] Through a simple experiment one can observe the productivity of polarity. When we look through a prism at the two diagrams of black and white juxtaposed squares in reverse positions what we see is

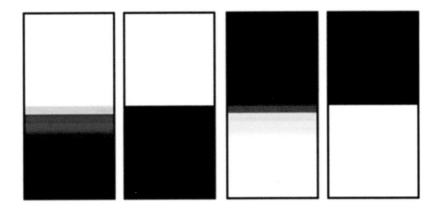

Fig. 24

Fig. 25

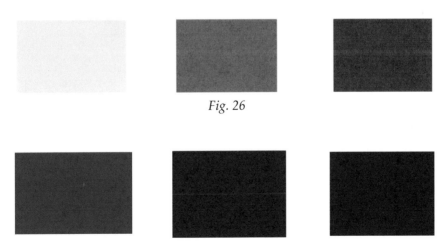

Fig. 26

Fig. 27

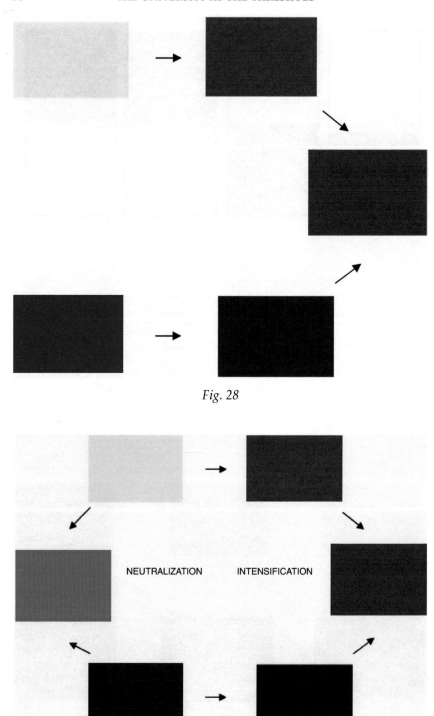

Fig. 28

NEUTRALIZATION   INTENSIFICATION

Fig. 29

presented next to each diagram; this is that the two halves of the spectrum are produced—the yellow-red part and the blue-purple part (see Fig. 24).

Through Goethean colour studies it can be shown in simple ways that the colour of the spectrum closest to light is yellow, and that the colour closest to darkness is blue. We can do this by looking at light and darkness through semi-transparent mediums. For example, if we look at our own wrist we can observe that through the semi-illuminated thin layer of the skin the veins close beneath the surface carrying dark venous blood appear blue (the lightening of the dark). When we put a sheet of paper in front of a light globe it appears yellow and red if the paper is thicker (the darkening of the light). This is related to observations Goethe made himself—namely, that the sky appears blue when the blackness of outer space is observed through a sun-illuminated admosphere and the brilliantly bright orb of the sun appears yellow at sunrise or sunset seen through the thickness of the atmosphere (which may also be dust- or smoke-filled in which case the sun will appear more red).

In such ways yellow and blue can be shown to be the 'colour representatives' of the absolute polarity of light and darkness (see Fig. 25).

This is confirmed by our experience of the colours themselves. If we participate in the colour yellow through cognitive feeling we perceive its radiant inner activity, carrying us outwardly in all directions. If we hold that gesture as an inward picture then we come to understand that this is the essential gesture of light. If we likewise participate in the colour blue, then we experience a negative, inwardly retreating gesture, moving into itself as it were, and grasp that this is the essential gesture of darkness.

To understand the creation of the colours we need to keep this polarity of yellow and blue clearly within ourselves as a dynamic picture formed of two opposing gestures. These gestures govern the whole realm of colour creatively; every colour is a creation of a specific relationship of these gestures. Let us consider orange: the radiant power of yellow is gathered and strengthened by the darkening force which brings this radiance to something of a centre. With red this tendency is heightened still further so that the inner activity of this colour appears as a force which is outward in direction (like yellow) yet is concentrated to a very great degree such that it appears to actually confront us.

Thus we see that the three colours yellow, orange and red reveal themselves as a metamorphosis expressed in degrees of *intensification* (see Fig. 26). We can place the three next to each other as a meaningful metamorphic sequence and permeate this sequence with cognitive feeling, backwards and forwards as necessary until the gestural language is legible. At the yellow end we 'lose ourself' in a radiating quality; at the red end the outward gesture becomes concentrated and we 'find ourselves' in the sense of being challenged by a forcefulness directed at us.

Now we can cognitively participate in the colour polar to yellow—blue—and the colours which develop from it metamorphically; to do this we again need to hold an inner picture of the interaction of the polar forces within the wholeness of the colour realm. Let us consider the colour indigo in which the inwardness and darkness of blue is intensified and lifted by the lightening force. In purple/violet this tendency is further intensified; now we find a gesture which still expresses the restless, negative quality (revealing a proximity to darkness) but which builds this quality into an inward richness and intensely living presence.

Again we see three colours—this time blue, indigo and purple/violet—reveal themselves as a metamorphosis expressed in degrees of intensification (see Fig. 27). Again we can place the three next to each other as a metamorphic sequence and permeate this sequence with cognitive feeling so that we learn to read the gestural language. In the blue we 'lose ourselves' in being drawn into an infinite depth. In the purple/violet the inward gesture becomes more concentrated and we are 'held' by this colour.

Thus there are two intensification progressions; one in relation to the yellow/red colours, the other in relation to the blue/purple colours. We can now place these two intensifications together and experience how both culminate in magenta. The inner gesture of magenta holds an essence equally of light and darkness and does so in a way which represents the intensification of both, coming from different directions (see Fig. 28). It has both the outwardly projected strength of red but the inward concentration and depth of purple. Significantly, this climax of intensification in the realm of colours is not part of the Newtonian 'spectrum of light' and magenta is thus not considered to be a colour.[128]

If we return to the polarities yellow and blue—the 'representatives' of light and darkness—and instead of intensifying them into magenta we simply combine them, green is produced. Entering into the gesture of this colour with cognitive feeling we experience the neutralization of the outward and inward gestures of yellow and blue (see Fig. 29).

This is how the circle of colours completes itself and reveals itself, through the language of gesture, to be a meaningful organization. Each colour is an image or expression of the whole, for each bears within it the essence of both light and darkness. Here we perceive the creative power and driving force of polarity, giving rise to a 'third thing'—the circle of colours—which still carries the opposing forces of polarity within it. The dynamic unity of the poles gives rise to an *organism of colour*, a differentiated wholeness, complete unto itself. Through metamorphosis driven by intensification, and through neutralization, the circle of colours comes into being in which every colour is organically related to every other. What is elementally expressed in this colour organism is immeasurably heightened in the organic wholeness of plant and animal forms and in the form of the human being.

Chapter 6
Exact Sensory Imagination, Delicate Empiricism
& The Law of Compensation of Parts

Intellectual Cognition (Philosophy)

None of the Goethean principles is separate from any other; we have seen that it is impossible to conceive of metamorphosis without an understanding of both polarity and intensification. In the same way, the cognitive capacities of exact sensory imagination and delicate empiricism are intimately connected to one another and the two together serve as the basis for the practice of Goethean science in all its aspects and dimensions.

The first thing which is likely to strike the contemporary student is that the expressions 'exact sensory imagination' and 'delicate empiricism' are rather unusual and even contradictory. It may seem extraordinary to juxtapose 'exact' with 'imagination'; also it may be surprising to hear about 'delicacy' in connection with the supposedly rigorous 'hard logic' and value-free methodology of conventional empirical science. It is the task of the philosophical aspect of these Goethean studies to clarify such questions.

The meaning of exact sensory imagination is closely linked with that of delicate empiricism, especially through the words 'sensory' and 'empiricism'. Empiricism is the view that knowledge of the world comes about primarily through sensory experience. Science, as is well known, is founded upon empiricism because it wants to build its theories on facts which come from actual observations of things rather than 'mere speculation' or 'innate ideas'. In fact, our experience and understanding of art depends just as much on empiricism, on perceptions which are mostly visual and aural, and just as little on 'mere speculation'.

Both science and art depend upon experience of the senses but, normally understood, there is a very big difference in the way this sensory experience is worked with. From the side of science, the question of empiricism goes far back into the history of philosophy, but ideas decisive for its role and meaning in modern science were set forth in recent centuries, in particular in the work of John Locke with his views on primary and secondary qualities. Primary qualities, he said, belong to the reality of the object itself (measurable attributes such as extension, motion, solidity and number) whereas qualities like colour, smell and sound belong to the interaction of the object and the human perceiver. Science took the approach of seeking its foundation on the measurable 'primary qualities' and relegating the secondary qualities to 'subjectivity'. In art, of course, sensory experience of colour and sound is everything.

While science proceeded on the basis of measurability, philosophy continued to ponder the separability of primary and secondary qualities. At the beginning of the eighteenth century George Berkeley presented the view that, if we strip an object of its secondary qualities, it becomes difficult to establish in a concrete way that there actually *is* an object. This thought was amplified by Kant who claimed that neither primary nor secondary qualities could be ascertained as belonging to 'the thing in itself'.[129] Taking his lead also from the philosopher David Hume, Kant recognized that the mind is *involved* in the perceptual knowledge of the world and doesn't work by merely making impressions of things, like a mirror. Kant went so far as to suggest that the mind actively *produces* the world, thereby opening the way to German idealism and the philosophies of Fichte, Schelling, Schlegel, Hegel and others.[130]

Such considerations led Kant to an investigation of the role of the imagination in knowing. He used the word 'imagination' to refer to the image-making aspect of the mind, to distinguish it from the logical intellect or reason. Only sometimes does imagination take the form of what we call 'fantasy' which is close to dreaming; in fact, imagination is inherent in what we mean by consciousness and what it is to know the world. Kant was specifically interested in imagination in relation to the way we make aesthetic judgements, through the working of the artist and in aesthetic appreciation. He understood imagination to be the way we judge an artistic work's integrity, the way it holds together as an organization of parts— its wholeness, in other words. In this way Kant sought to avoid the pitfalls of relativism and solipsism in the area of epistemology.[131]

Kant made a clear distinction between a *concept* and an *image*. Science, he considered, has to do with concepts—that is, with *thinking*—and aesthetic judgements with *imagination* through which we develop aesthetic *feelings* in relation to the objects of our study. Thus he divided the human being between the scientist and the artist—and this, we might add, remains the normal view today. The genius, he says, must be original, and 'cannot itself describe or indicate scientifically how it brings its product into being'. In this way he sharply distinguishes between the artistic genius and the scientist; there can be no 'scientific genius' for Kant.[132] Likewise, science cannot describe conceptually how an organism brings itself into being; Kant recognized that an organism is self-creative, the cause and effect of itself. He wrote that 'there will never be a Newton for a blade of grass', meaning that organic form cannot be grasped through the logic of cause and effect relations.[133]

The philosopher Friedrich Schelling, in part because of his close connection with Goethe, brought it about that this division between the scientist and the artist was greatly diminished—at least in philosophical terms. This was possible because he broadened Kant's understanding of what an *idea* is, to what he called the 'living ideas' of nature. Schelling realized that not just particular organisms but nature as a whole is a living creative being. His view is that the mind is not limited in the way Kant believed because human experience tells us that the creative ideas of nature *can* be grasped. The point he makes is that the comprehension of living, creative ideas is natural to the artist who *thinks creatively* through the active imagination; this is so because artistic creativity *is* nature's creativity lifted to a higher level of potency. Art is the special place in human life where the creative forces in all of nature come to the surface, as it were, and become something conscious.[134] When we study and understand works of art, he writes that it is 'the spiritual eye [the eye of the imagination] that penetrates their husk, and feels the force at work within them'.[135] Likewise, when we study works of nature the eye of the cognitive imagination can grasp the living, creative forces (ideas), and this should be the work of the natural scientist. For this reason Schelling aimed to inaugurate a new poetics of nature in which the aesthetic and the scientific would be united.[136]

A deep reflection on the role of the imagination is also found in the writings of two English poets—Coleridge and Wordsworth—with Coleridge's theories of the imagination owing a great deal to Kant and Schelling. Coleridge's 'primary imagination' relates to what we all do naturally when we perceive and understand the world, his 'secondary imagination' to the special creative work of the artist. Wordsworth, like Schelling, understood how the creative imagination can enter into nature and perceive the living, creative forces at work. He expressed this in these well-known lines: '...with an eye made quiet by the power/Of harmony and the deep power of joy,/We see into the life of things'.[137] Still, in this account of the power of cognitive imagination in the work of the artist, we remain in the realm of philosophy and gain no insight into how imagination can enter into the actual practice of nature study.

Goethe conceived organic wholeness as a kind of economy of living, formative force, as 'a budget (*Etat*) in which . . . the main sum (*Hauptsumme*) remains the same, for if too much has been given on one side, it subtracts it from the other side and balances out in no uncertain manner'.[138] His idea was inspired by the thinking of Kant on the nature of the organism and the ideas of the biologist Blumenbach on organic formative drive (*Bildungstrieb*). Kant had written in his *Critique of Judgement* that organisms possess 'a self-propagating formative power (*bildende Kraft*) that cannot be explained by the capacity of movement alone, that is to say, by mechanism'.[139] In relation to the constitution of parts or organs of an organism, whatever is emphasized formatively in one part or group of parts must lead to a corresponding diminishment in the others according the principle of 'giving and taking'. This is what Goethe referred to as 'the law of compensation of parts'; when one organ has 'preponderance'

(*Übergewicht*) it is creatively compensated for in the overall constitution of the organism.[140] This is not a logical or intellectual formulation relating to lawful mechanical-causative processes in nature although on that assumption it was refuted by Charles Darwin.[141] Rather, it is an insight into the creative or formative process of living form; the lawfulness Goethe is speaking of can only be grasped by the eye of the imagination which works from an aesthetic sense of the wholeness of form and is able to make judgements about the distribution of parts and overall balancing of the parts within the whole.

Goethe called the kind of scientific thinking which is founded on such aesthetic judgement an 'exact sensory imagination'. With this simple expression he makes explicit the relation between science and art in his phenomenological way of working—the method is scientifically exact yet artistically imaginative. He makes the further significant point that science is learning the way of exact sensory imagination from the arts which are 'unthinkable' without it.[142] What stands as a *possibility* in relation to the activity of science is something which is part and parcel of what art means in the most fundamental sense. In terms of conventional and recognized practice, science is yet to enter cognitively into the realm of living creative forces in nature; and it is in this realm of understanding that art has so much to offer.

From the foregoing we can also come to appreciate what Goethe meant by a closely related expression—'delicate empiricism'. As we have noted, the empirical approach is fundamental to both scientific and artistic practice. Goethe is suggesting by this expression that there can be different *kinds* of observations of natural form, that observations can be carried out with different *qualities* of attention. In his essay *The Experiment as Mediator between Subject and Object* he discusses how the history of science teaches us the difficulty of scientific observers renouncing their own point of view in experimental work, how desire and dislike very easily enter into the process of observation. He writes in this essay that true scientists 'must find the measure for what they learn, the data for judgement, not in themselves but in the sphere of what they observe'.[143] Science in this way of thinking is a form of renunciation, a giving of oneself for the sake of the truth; the scientist approaches and treats the phenomenon with the greatest delicacy and respect, even with reverence. This happens when the scientific mind moves beyond its objective detachment to involve itself with the phenomenon imaginatively.

This is the point which has already been alluded to in Chapter 1, under the heading of *care*. We can care for another being in the way we go about observing it; care in empirical research means openness to the form of something as immediately experienced through the senses, to its naked existence yielding itself towards us as researchers. Comprehending the living idea of the things of nature will not take place through some kind of mechanical experimental procedure; it occurs through the way in which the 'eye of the imagination' is able to enter with cognitive feeling into the form of the thing it is studying and experience it from within. Only a cognitive feeling is capable of such sensitive, 'delicate' participation in the form of another being; as the Goethean scientist Jeremy Naydler expresses it: '… the essential plasticity of the act of observation means that it can be shaped and guided by the mind's attentiveness to what the phenomena are really saying'.[144]

Goethe cautions us by suggesting that a delicate, caring empiricism is something immensely difficult to achieve, even while being a vital necessity in our time. He writes:

> There is a delicate empiricism which makes itself utterly identical with the object, thereby becoming true theory. But this enhancement of our mental powers belongs to a highly evolved age.[145]

It may indeed be true that such enhancement belongs to a time which is still to come, at least in the sense of a broadly recognized practiced scientific methodology. But that doesn't mean it is impossible in the here and now to initiate in one's scientific practice what appears to be a necessity and a good. Both exact sensory imagination and delicate empiricism are at early stages of development within the realm of the sciences and to a large extent the humanities, and it is for this very reason that they need now to be introduced and developed at the tertiary level of education.

Cognitive Feeling (Goethean Science)

a) Delicate empiricism and exact sensory imagination

It hasn't always been the case that the methods of empiricism were taught and practiced within the context of universities. In the medieval universities the work of students was entirely devoted to absorbing key traditional texts, in law, medicine and other subjects. Indeed, science education today— in both schools and universities—is not so far distant from the medieval methods insofar as it promotes experimental work as simply the means to confirm an already learned and accepted theory.[146] Empiricism, ideally, pushes the student towards the coalface of learning—towards developing powers of perception and individual responsibility for knowledge.

The Goethean approach is not one which takes sensory experience as merely evidence to support an abstract theory. Goethe's injunction was: 'Let us not seek for something behind the phenomena—they themselves are the theory.'[147] In the Goethean study of an organic form we learn to value each and every sensory experience in and of itself as the eye of imagination begins to perceive the place of the particular within the whole. Words other than 'delicate' could be used: we could just as well speak of a 'respectful empiricism,' a 'caring empiricism', a 'sensitive empiricism'. How then do we proceed with the practice of exact imagination and delicate empiricism? This has already been introduced in Chapter 2 in relation to Goethe's comparative method. Goethe said that art is unthinkable without exact sensory imagination and it is in the realm of art that the apprenticeship in this capacity is undertaken. The activity of understanding art is not theoretical but phenomenological—we try to *read* the work through involving ourselves thinkingly in the phenomenon. Exactitude belongs to artistic judgement, both in artistic practice and in interpretation, even though it doesn't involve measurement.[148]

Let us look at the following famously enigmatic work, *Melencolia 1* by the German Renaissance artist Albrecht Dürer.

Fig. 30: Albrecht Dürer, Melencolia 1, *engraving, 1514.*

This image is no simple representation of a familiar object; indeed, its enigmatic quality is partly what moves us to endeavour to understand it. We direct our eye with great care over the whole surface many times, connecting each object with every other object in the whole optical field of the work 'in unbroken succession'—the rainbow, the ocean, the ladder, the human and geometric figures, the bells, scales, 'magic square' and hourglass, the tools and millstone lying around. Everything is quiet; nothing is happening. We are *delicate* in our perceiving because we are respectful of the work, wishing just to take it in and not jump to conclusions about it. We are *naturally* thus delicate in our approach to things which we love and respect; this attitude of profound respectful receptivity is our starting point.

Delicate empiricism gives rise to exact sensory imagination. Through our delicate perceiving we seek to grasp the work in its wholeness, its unity, because only from out of this unity can any particular part be

Fig. 31

understood—for example, the geometrical figures and the angel. Each element is held in the imagination exactly in relation to each other element, thereby resisting a too-speedy intellectual interpretation being foisted on it, never assuming that anything has been arbitrarily placed or shaped within the work. Such is the method of exact sensory imagination employed in artistic interpretation.

We may now focus for a while on the figure of the angel (see Fig. 31). Nothing here is measureable—that is, exact in the mathematical sense. Yet our *attitude* and *approach* is of wishing to comprehend *exactly* what we perceive. The darkened face and arching brows reveals eyes which stare fixedly forward; the head is bent forward and is supported by the bent arm in the gesture of a thinker. This is no joyful figure in spite of the angelic wings; the expression is one of brooding or deep reflection. The figure 'speaks' to us more readily than other elements of the work; the physiognomy and language of bodily gesture show how precisely shaped the figure is in relation to the whole work. Even a relatively minute difference—particularly in the facial expression—may change our reading.

Let us now focus our attention on three contiguous forms which have no obvious meaningful relationship to each other (see Fig. 32). This configuration also speaks in an expressive language or physiognomy of form. These forms have been placed in a particular way by the artist, in a vertical alignment: below is a small, pure, bright sphere; in the middle is a curled up organic animal

Fig. 32

form in an attitude of unconscious repose; above is a larger polyhedral form. Deciphering the meaning of these forms is not what is of immediate importance here; what is crucial is how we go about perceiving them. Only through exact sensory imagination can we learn to *approach* the hidden meaning of which the visual language of these forms is speaking. We observe them at length, sensing in many different ways their relationships with each other and then with the rest of the forms in the work, not jumping to conclusions. We might, in honesty, have to say that we do not understand these forms and admit the need to return to them, again and again.

A first step has been taken; we have begun to develop the 'organ' of exact imagination necessary for a science of the living world, a qualitative science. We have practiced exact sensory imagination in relation to a black and while tonal artwork and if we turn to coloured works the requirement of exactitude is the same. In the *Sistine Madonna* by Raphael, the blue colour of the robe stands out as central within the optical field of the work. Again, we spend time with delicate empiricism, respectfully taking in every part of the work and this includes the different colours in relation to the total composition. Looking at the work with 'fresh eyes' we perceive that the colour of the robe bears relationships with every other element within the wholeness of the work—to the duller tones of yellow, green and red, to the white ethereal space in which the blue is central, to the surrounding figures and to the figure of Mary herself in that the blue robe shines forth with an almost heavenly purity and nobility. Now we can further develop our exact sensory imagination of that colour: we experience the inward-moving, receding gesture of the blue of this robe, its solemn depth, and can connect it through exact imagination with the baby Jesus and Mary who are calling us to deep spiritual reflection. In the field of meaning of the work there is an exactitude about the artist's use of this colour in this particular central position within the work. In a work of art we must conceive colour in this qualitative way; it would be unfruitful to talk merely about the wavelength of blue in this context.

Turning now back to the living forms of nature, we have through our art study gained experience in an approach which will allow us to perceive colour—also such qualities as scent and sound—in an exact imaginative way. This approach is not 'merely artistic'—blue is the same phenomenon wherever we find it, in a flower as much as in a painting. If we choose to study a plant which has a blue flower it becomes necessary to study the meaning of this flower in terms of the whole organism. Just as the blue robe is an aspect of the whole work of art, so it certainly is the case that the blueness of the flower is an aspect of the wholeness of the organism.

Fig. 33: Sistine Madonna, *oil painting by Raphael.*

Fig. 34: Salvia officinalis (Labiateae).

Let us look, by way of example, at a species of salvia (common sage) which has a blue flower (Fig. 34). Here we are focussing on the floral form, but to study the plant fully it would be necessary to consider the plant in detail and as a whole. In whole-plant Goethean observation work we proceed just as we did with the works of art—through delicate empiricism we move with our eye (and senses of touch, taste and smell) around and through the entire plant, respectfully, caringly, not assuming that any part, form or quality is arbitrary or meaningless within the whole.

This work may lead on to a comprehension of the plant through exact sensory imagination. In terms of floral form and colouration we can say: in the blue of the petals, just as with the robe of the *Sistine Madonna*, we experience an inward-moving, receding gesture, a solemn depth, as opposed to, say, the experience of yellow which has an energetic gesture of outward radiation. In the salvia the floral blue can easily be correlated with the physical form of the flower—a funnel shape. On the level of gesture the shape and the colour are one and the same thing: inwardly receding, self-enclosing. This flower is not of the type where the petals are outstretched and the central organs (the pistil) and stamens prominent. On the contrary, it is a tubular enclosing form with its central organs hidden deep within. The correlation between the gesture of the colour and the floral form is no coincidence; indeed, this correlation holds good to a large extent with all blue-coloured flowers.[149] Nevertheless, as with the painting by Raphael, the floral colour and form must be seen within an exact imagination of the whole and if these two do not immediately correlate a meaning must be found through a more detailed and encompassing study.

One other element may be mentioned here: the fact that its oblong leaves have a slightly peppery flavour. Scents and tastes also have characteristic gestures and about peppery flavour we can say, in general terms, that it is neither sweet (elevating) nor sharp (contracting). We bring the three qualities—the blue colour, the funnel shape and the peppery flavour—together in the exact imagination, testing them at length in relation to each other. The correlation of floral colour, form and leaf flavour may not be so easy but we might tentatively say that they have all to do with the same 'inwardly receding' gesture. From this it is a relatively short step to seeing that *every* part of a plant (and indeed every form within the realm of nature) is expressive of a particular gesture, that the whole of nature can be 'read' as a language of gesture. This is the fruit of our approach, moving from delicate empiricism to exact sensory imagination. As with the three vertically arranged forms in *Melencolia 1*, we might have to admit the need to return again and again to this mere glimpse of 'who' this plant is. These indications concerning salvia are really only the beginnings of what would be needed to become a comprehensive study, if one is to do justice to the methods of a goetheanistic life science.

Fig. 35: Two white oaks (Quercus alba) growing 200 metres apart, one in an open field and the other at the forest edge.

Goethe identified two fundamentally opposing gestures in the general form of plant growth, from seed to flower and to the formation of new seeds—expansion and contraction (see also Chapter 7). We read these gestures with a form-sense related to the cognitive feeling through which we grasp the different gestures of colours. In leaf form and in overall habit plants express the gestures of expansion or contraction or degrees between. Some plants tend to highly contracted habits—for example, those growing near the polar regions, such as the sulphur buttercup. In other regions plants sprawl over the earth expansively. Plants of the same species, depending on the environment in which they are growing, may take an expansive or contractive form. This is something noted by Goethe when he crossed the Italian Alps in 1786; he observed that the willow, gentian and rushes growing in the higher altitudes tended to be taller and thinner than those growing on the plains.[150] Such plasticity can be seen in plants of all kinds, even those growing in very close proximity but in different orientations to the Sun (see Fig. 35).

Some species even display a plasticity of form *in the same individual*; for example, *Cabomba* sp. (see Fig. 36). The leaves of this plant which float on the surface of the water have a rounded, expanded form whereas those beneath the water's surface are linear and root-like.[151] This shows us that our whole-organism study through exact sensory imagination must take in the qualities of the environment, extending to the existence of the plant between the polarities of Earth and Sun.[152] In the case of this plant, cognitive feeling grasps the spreading, rounded gestures of the top leaves in terms of the expansive quality of the water's surface and the linear gesture of the lower leaves in terms of the enclosed mineral-rich body of the water, the normal domain of the roots. Here leaves become root-like in form; root process is 'pushed up' into the realm of the leaf.

More than anything, such examples speak of the need for a truly delicate empirical approach to the study of living things. We perceive

Fig. 36: Cabomba sp.

the flexibility and creativity of organisms and realize that our approach must be equally flexible and creative. From a related direction we can consider the plasticity of plants within one family in terms of an emphasis on flower, stem, leaf or root forms. In the case of the cabbage family, Brassicaceae, we start with delicate empiricism; carefully we take in the particular qualities and forms of the different species. We run through and around all the parts, building up a vivid picture of each. Gradually this can transform into the kind of comprehension we call exact sensory imagination: imaginatively we 'sculpt' the form of one plant and transform it into another systematically throughout all these manifestations of the cabbage family. The flower swells to become the cauliflower and broccoli, the leaf swells to become the cabbage, the stem to become the kohlrabi, and root to become the swede. In this way we enter into the language of gesture as it pertains to this plant family.

← Rutabaga or swede, a swollen root

Clockwise from the swede:
- Brussel sprouts or swollen buds
- kohlrabi or swollen stem
- cabbage or swollen leaves
- cauliflower or swollen flower

Fig. 37: Members of the cabbage family.

Turning now to the human sphere, this time to the study of human history: it was noted in Chapter 3 that Wilhelm von Humboldt applied a Goethean methodology to this realm of historical phenomena. Human history is a manifestation of life—*human* life—and must be studied through a living methodology with delicate empiricism and exact sensory imagination. Humboldt speaks of the need for artistic breadth of vision and imaginative delicacy (he calls it a 'subtlety and tact') required to read the language of historical phenomena without destroying their 'simple and living truth'.[153] This, it must be said, represents an advanced form of Goethean study which can be prepared for through the study of artworks and forms in nature.

Humboldt explains that to study the events of history we must begin with an 'exact, impartial, critical investigation of events'; this represents for him the 'exact' aspect of his method. In this he 'subordinates his imagination to experience and the investigation of reality' and thus the imagination cannot just impose its own invented forms on reality.[154] In other words, we proceed through delicate empiricism toward exact sensory imagination. Humboldt presents this study of the events of history as taking place within a nexus of activities and forces which includes soil and climate, the intellectual character of nations, the particular character of individuals, the influences of the arts and spiritual inspirations.[155] With such material produced by the inquiring intellect, the eye of the imagination develops a picture of events by dwelling in them feelingly—that is, through cognitive feeling. This deeper perceiving with the eye of the imagination is what he calls 'the intuitive' or 'connective capacity', the power to grasp manifestations of an event which may be 'scattered, disjointed, isolated' so that every aspect gradually appears as a manifestation of the whole. Humboldt writes that the historian must become active, creative in the sense that the creative powers of the historian become an 'organ' for perceiving the otherwise hidden 'creative forces' at work in history.[156]

For teaching purposes, any example can be taken and worked with; a prominent event such as the French Revolution is suitable. The following can act as a guide for the work of exact sensory imagination (in the case of historical studies one relies on the empirical studies of others), building on the factual as far as it is known. Here only an indication of the method is possible but a fuller account of this historical event using a goetheanistic method can be found elsewhere.[157] What is important for teaching purposes in an orientation course of study is the approach, not the depth of knowledge of any particular aspect.

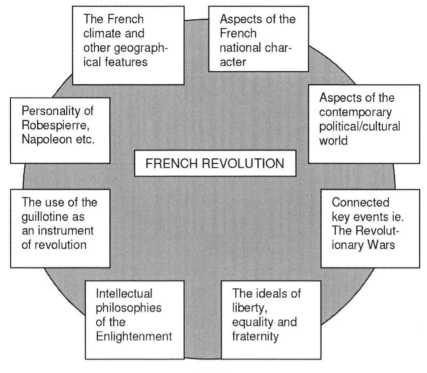

Fig. 38

The intellectual-mechanistic approach to history talks mainly about the causes of the French Revolution, the approach which Humboldt describes as the 'mechanical determination of one event by another'.[158] In the goetheanistic approach developed by Humboldt, through delicate empiricism to exact sensory imagination, one enters receptively into these different 'parts' or aspects, not seeking out defining causes but gradually allowing a vivid imaginative picture to form by inwardly 'testing' the nature of each aspect in relation to all the others. The climate, the individualities involved, the cultural context—all these have a quality which can be 'read' in relation to every other, in exactly the same way we learned to read the gestural language of the parts of a work of art or a plant in the examples given above.

A medical symptom is a manifestation which speaks of processes which are hidden from immediate view; in a related way the reading and interpreting of historical events can be called an historical symptomatology.[159] The French Revolution is a particularly significant event in the history of Western European civilization but as a *symptom* it speaks of a moment or phase in the living, evolving consciousness of humanity as a whole. It is what Goethe called a 'pregnant point' (see Chapter 7) which, in relation to historical research, means an event on the surface of life expressive of certain forces at work everywhere but which do not show themselves otherwise in such a significant manifestation.[160] Just as in Goethean phenomenological plant or animal studies we may obtain a glimpse of 'who' a plant or animal is, of the living idea of that organism, so a symptomatological historical study may lead to insight into the particular creative powers at work in human evolution in a particular age.

What, then, are the 'creative forces' at work in European civilization, one of whose most significant 'symptoms' was the French Revolution? The Revolution was characterized by its 'untold confusion', with the French populace motivated by ideals of liberty, equality and fraternity which were scarcely understood and which were not able to find harmonious social embodiment at that time. Instead these ideals took the form of slogans which were pressed haphazardly upon human nature: the human being shall be free, equal, and live on Earth fraternally![161] Rudolf Steiner, giving expression to a symptomatological account of the French Revolution, describes it as a violent beginning of what was seeking to realize itself in human physical, soul and spiritual life—at that time, still in our own and into the future. Forces at work beneath the surface of human life had long prepared for this event: *freedom* seeks realization in the human sentient life (what Steiner calls the soul) which is the meaning of human individualization. It has gradually come about that human beings consider themselves to be free, autonomous selves. But individualization is not just about freedom; free human individuals may choose to associate and work together in ways completely different from the kinds of communities or work which occurred through the older instinctive consciousness. That is the meaning of *fraternity*. And the dawning of human individuality is also the dawning of the sense of rights which belong to every individual *equally*, beyond all castes or class structures.

If we look back even further into history with our 'connective capacity'—that is, through the eye of the imagination—we find what might be called the 'prophetic forms' of what eventually became the significant event we call the French Revolution. For example, Theodorus of Cyrene (c. 300 BC) is famed for his teaching that the goal of life is happiness and that this depends on knowledge, not on reliance on the gods. The gradual liberation of the human inner life from the dominion of the gods through Western civilization can be followed imaginatively, so that each event can be perceived as a symptom *of the same force*. Thus we see that this kind of exact imaginative history research can open a window on what it means to be a human being of body, soul and spirit through which the students find *themselves*. In Chapter 8 we will follow these creative, formative processes further to see how individualizing forces in civilization—expressed through violence and through the production of the slogans of liberty, equality and fraternity in the French Revolution—are realized as a consciously created threefold social organization.

b) The law of compensation of parts

Turning now to another key 'letter' in the language of living form, we can consider what Goethe called the law of compensation of parts. In any living whole—whether it be a work of art or an organism—some aspects may be emphasized or enhanced, sometimes greatly, while others de-emphasized or diminished to the same extent. What we discover in all cases is a dynamic equilibrium or balance of parts within the whole. In teaching practice this scientific-aesthetic appreciation of living form can be cultivated in the first place through artistic experience.

For example, we may look at the Brahms's *Variations on a Theme by Robert Schumann*, Op. 9. Here the theme is stated through a strong balance of treble and bass parts which largely follow the same melodic and harmonic pattern.

Fig. 39: Excerpts from Variations on a Theme by Robert Schumann *by Brahms.*

In the variation the treble is reduced to a succession of arpeggiated or broken chords and the bass is not present at all for the first two bars. The bass then assumes the form of rapidly alternating semiquavers between the broken chord. So *emphasized* or strongly stated are the broken chords in the treble that the base is able to be diminished to nothing at all in the first

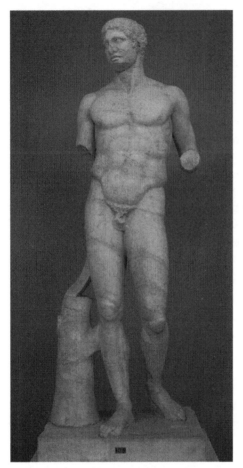

Fig. 40: A marble copy of Doryphoros by
Polykleitos.

two bars, then just to flourishes between the chords. But nevertheless—musically speaking—the 'budget' of forces which is expressed in a balanced way in the theme is maintained even in the onesided expression of the variation.

Now a work of Greek figurative sculpture can be considered in which a harmony or balance of parts is clearly evident (see Fig. 40). This is described as the *contrapposto* stance, typical of Greek art from the Classical period. The whole meaning of this work comes through this overall balance; nothing is exaggerated or overemphasized. This is an image of the ideal human being, brought forth from a profound spiritual sense of what is divine and beautiful behind or within every actual, living human being. It is an expression of what might be called 'spiritual fullness'.

By way of contrast, a work by the twentieth-century Swiss artist Alberto Giacometti can now be considered. Here the rigid verticality of the figure is emphasized to a great degree, suggesting the modern existential state of human beings who merely walk erect yet are devoid of the spirit and soul which can give meaning to this upright posture. To the extent that this nihilistic verticality is exaggerated, the peripheral features of the figure are diminished: the arms are reduced to stumps, the feet to mere enlargements, the features of the head—the ears, nose, lips—to generalized lumps. The elements of the body through which the human being extends out and relates to the world (the senses and the arms) are almost absent and each individual figure appears condemned, trapped, in separate physical existence.

Goethe provided various examples of what he meant by the 'law of compensation of parts' in relation to the animal kingdom. With regard to the giraffe he wrote: 'The neck and extremities of a giraffe

Fig. 41: Alberto Giacometti, Woman of Venice VII, *bronze, 1956,*
Art Gallery of New South Wales.

are encouraged to develop at the cost of the giraffe's body . . '.[162] The
giraffe provides perhaps the clearest example possible; the extreme
lengthening of the neck and legs (and also the head and tongue etc.) is
balanced by a correlative diminishment of the trunk. In a certain sense
the trunk—which is so evidently defined in an animal such as a cow or
horse—has 'disappeared' in the giraffe and appears as a continuation
of the neck (see Fig. 42).

Goethe also pointed to the inverse relationship between the dentition and the presence of horns in certain mammals.[163] What he is saying is that an animal with a full set of teeth in its upper jaw (such as a
lion with its upper incisors and large canine teeth) *cannot* have horns—
whereas a sheep, for example, has neither upper incisors nor canines
and has horns. As philosopher Henri Bortoft notes, such a correlation
between teeth and horns in a mammal would be considered merely
contingent to biology in the Darwinian paradigm.[164] What is certainly

Fig. 42

the case is that an appreciation of the necessity of this relationship can only emerge from a comprehensive 'whole animal' study through exact sensory imagination.

The compensatory correlation between horns and teeth has been explicated to a considerable degree by Goethean scientist Wolfgang Schad who shows that it is only meaningful when seen within encompassing patterns of formative activity in these animals. The ungulates (hoofed mammals) are characterized by protuberances on the head and this is connected with the overall physical organization of these animals in which the anterior (frontal) part is more or less emphasized—especially so with ruminants such as the bison. This anterior emphasis is an expression of a highly emphasized metabolic system and correlated with this emphasis is both the de-emphasis of certain teeth and the formation of horns. Shad demonstrates that the teeth-horn inverse correlation relates to the position of the horns on the skull, with the cow's horns situated at the back of the skull and positively correlated with the highly emphasized molars (and de-emphasis of incisors and canines).[165]

Fig. 43: A bison (the European bison, Bison bonasus) showing anterior emphasis of body form.

If we refer to the 'law' of compensation we must be careful to make clear that this is not a law in the sense meant by physics and biology when they speak of 'physical laws'. What we are dealing with here is the lawfulness of the animal *typus* which doesn't stand as an explanation 'above' all particular forms, but has 'flowed into' the particular living form, creating it. The difference between the natural law and the lawfulness of the *typus* has been discussed in Chapter 2. To comprehend the lawfulness of 'compensation of parts', of an organism's 'budget' of formative forces, it is necessary for thinking to re-enact the creative activity of the *typus*, and this is exactly what exact sensory imagination leads to. Through exact observations on the physical level (delicate empiricism) we unfold exact feeling insight into the living form, which can also be called cognitive imagination. Therefrom may develop an understanding of the precise formative gestures at work in the organism, which is the fruit of cognitive will.

Cognitive Will (Artistic Practice)

Observational drawing:

This work is like a foundational practice and preparation for any other form of artistic practice. The focus can be any form, familiar or unfamiliar, natural or human, and this form might be chosen because it will be part of a more developed study. The intention of this work is to observe the phenomenon with 'new eyes', taking it in in great detail and not assuming that any particular part or aspect is unimportant. In relation to the practice of exact sensory imagination, this observational drawing concentrates on the 'exact' and 'sensory' aspects. Although many different styles are possible—through the use of fine pencil or pen for example—the approach can be the same in all circumstances: this is to scrutinize the object, not just in a cold methodical manner, but with great care and respect for its physical form which is yielded up to the observing eye; in other words, through a delicate empiricism. An important extension of this work is to carry out such observational drawing a day or even a week after the careful observations were first made, to develop the faculty of exact memorization.

Colour conversations:

This practice heightens one's imaginative understanding of colour gesture and also develops the sense of placement within the field of a work of art, in this case a sheet of paper or a canvas. It can be carried out with two or even three people working with a range of prepared colours. One pure colour is applied by one person, in some shape or form, in one part of the blank surface; even this first step should not be rushed but take place through a feeling-sensing of the exact quality of this colour in relation to the space of the empty surface. In a slow and deliberate manner, continuously sensing the whole changing field of the painting in relation to the specific gestures of the colours, one colour after another is applied by each person sequentially, in whatever way seems appropriate. Each decision is based on a 'conversation' between contiguous colours—conversation in the sense that, for example, yellow with its radiating pure gesture comes into a certain dynamic relationship with purple, with its more inward richness. Clearly, every resultant 'work' will be different depending on the individuals involved, yet the exercise is founded on exact perceptions of the qualities of the colours and how colours interact.

Poetical animal gesture study:

Any animal can be chosen and it is necessary to have some familiarity with the natural habitat of this animal. For this exercise photographic images of the animal can be worked with. The animal is carefully perceived as a whole and each part (and this includes shape, colour, behaviour, cry) is looked at in terms of the whole form, just as we do with a work of art. This should be a lengthy process. The 'poetical' aspect of this study is finding words and phrases which capture the gestural quality of each of these parts and their relationships; it is not necessary to write this in the form of a poem; the main aim is to find a suitable expressive language for the 'language' of gesture which is the animal form.

The compensation of parts:

This exercise develops the purely artistic sense of form and composition within a fixed visual space or frame. Working with compressed charcoal or a thick graphite stick and an eraser, an initial shape of any kind is developed in some part of a sheet of paper; this could perhaps be a roughly circular form. All forms should be filled shapes, not outlines. From this initial form other forms are developed as extensions, or perhaps isolated shapes in other parts of the defined area, in ways which balance or 'compensate for' the initial form. As the enhancement of the initial form takes place a corresponding diminishment of other areas is carried out, using the eraser. Nothing here is predetermined—everything is worked out continuously through the artistic sense of composition. The process should continue until a satisfactory composition results.

Case Study: Understanding The Australian Platypus Through Exact Sensory Imagination and Delicate Empiricism

Background to this Goethean study:

What follows is only a preliminary exploration, intended to show a way of studying natural phenomena with delicate empiricism and exact sensory imagination. It is the kind of study which is suitable for students working within an orientation course and could be undertaken jointly by students and teachers.

In Australia there are 357 indigenous species of mammals. This list of native mammals includes 5 monotremes, 159 marsupials and 64 rodents (the young of monotremes are born from eggs; marsupials suckle their young in a pouch). No monotremes are found on any other continent and only one marsupial exists elsewhere—the American opossum. Just these facts alone are sufficient to draw the enquiring mind quickly to the question as to why these special mammalian forms are concentrated on this landmass. We take up this question here in a 'delicate empirical' consideration of the unique features of the Australian landscape and one of its mammals, the monotreme called the platypus.

This question of the specialized mammalian environment of Australia has had for a long time a chief point of focus in the enigma of the so-called duck-billed platypus (*Ornithorhynchus anatinus*), one of the five monotreme species (the other four being species of echidna or 'spiny anteater', found only in Australia and New Guinea). Skins of the platypus had been sent back to England for study as early as 1797 and in 1802 the animal was dissected, showing (as with the echidna) it had a common orifice or cloaca for excretion and reproduction, as do birds, amphibians and reptiles.[166] Writes one contemporary expert on the subject, Professor Tom Kemp:

> The greatest mystery of all concerning mammalian evolution stretches back for 200 years: the question of what exactly the monotreme mammals are, and how they relate phylogenetically to therians [ie. all other mammalian forms].[167]

In recent times a concerted effort has been made to solve the 'mystery' of the platypus: it was specially selected for a genome-mapping research project involving around a hundred scientists worldwide.[168]

For Darwin himself the case of the platypus first appeared as the most felicitous evidence for his theory of evolution by natural selection, where 'missing links' between known species must exist. In his *Origin of Species* he called the platypus a 'living fossil'—a remnant of the otherwise extinct forms of animal that represents a transitional form between egg-laying reptiles, birds and milk-suckling mammals.[169] The platypus undoubtedly can be considered a key phenomenon in the history of biological science—however, for a reason entirely different from the fact that it represents a fascinating intellectual conundrum. It is key because it points clearly, perhaps better than any other animal, to the *impossibility* of separating facts or data from the whole picture—and that certainly includes genetic data—and then pressing these into the service of an intellectually-derived theory which stands outside the sphere of life itself. The scientific enigma of the platypus highlights the imperative of gaining a genuinely living understanding of the whole animal through delicate empiricism and exact sensory imagination.

Conventional scientific methodology uses factual knowledge to explain the animal in terms of logic-derived theories, of treating living things as puzzles to be solved, instead of asking: what thinking approach is requisite 'if we are to live up to the object, be on a level with it?'[170] This means working from our own observations, together with what we might learn from the investigations of others, in the mode of a delicate empiricism—seeking not only to build a picture of *what* the animal is but *who* it is. *That* is the great mystery of the platypus, not at what point it branched off the tree of life according to a computer generated clade. If we are able eventually to progress some distance along this pathway of honouring the animal as a unique being—*then* it might be possible to fathom something of how it came into existence through evolutionary creative processes.

The platypus—a picture:

Through the method of delicate empiricism we begin to build an inner picture of this animal, simply receiving the facts 'delicately' into our thinking without trying to force them into a preconceived schema or theory, just permeating them with cognitive feeling which Goethe described as 'making itself utterly identical with the object'. At the same time we do well to keep in mind his further comment that 'this enhancement of our mental powers belongs to a highly evolved age'; this helps us to see that our task to properly understand the platypus is not an easy one. On the contrary, the Goethean approach to understanding animal life is highly demanding—but not in the intellectual sense. To build a vivid, living inner picturing of this animal we mentally gather up each fact or observation, our own and those of others, entering into each to find its inner gesture or tendency, sensing and testing this tendency against the qualities and gestural shaping of the picture which we have already built; this, as discussed in Chapter 2, is the comparative method. In this way the animal by degrees 'comes alive' in our thinking and meaningful patterns begin to emerge if this work is maintained and developed. What is set out below at least goes some way in this direction.

Fig. 44

The semi-aquatic platypus is found along most of the eastern coastal areas of the Australian mainland and in Tasmania. It inhabits small streams and rivers over an extensive range from the cold highlands of Tasmania and the Australian Alps to the tropical rainforests of coastal Queensland. The average length is 43 cm and the body and the broad, flat tail of the platypus are covered with dense, brown fur that traps a layer of insulating air to keep the animal warm. In a river they are difficult to distinguish from rocks and the water. All four feet are webbed but only the front ones are used for propulsion—the rear two, with the tail, for steering.

The animal is shy and elusive in nature, compared to the more aggressive water rat (*Hydromys chrysogaster*), a rodent, which is a similar size and is native to the same pools and streams as the platypus. The platypus spends most of its time submerged in the water, coming up for breaths approximately every minute so that it dives a thousand times in a 12-hour period. It is mainly nocturnal or crepuscular (appearing at dawn or dusk); it is largely solitary and territorial although these territories can overlap—with the males always dominant over females.[171] In overall shape the head connects to the trunk without a visible neck and the posterior part of the body is emphasized.

The platypus has a rubbery 'bill' similar in shape to that of a duck, but—unlike the duck—the mouth is not between two opening parts but is on the underside of the bill. The bill itself is used for digging and foraging extremely energetically and incessantly on the bed of rivers and for digging burrows; it is also a sense organ and has over its surface a very large number of electroreceptors and mechanoreceptors. It makes characteristic sweeping movements with its bill as it forages; this is how it moves and seeks its prey. When diving it uses neither vision, hearing nor smell; when submerged a fold of skin covers the eyes and ears. It is a carnivore, feeding on shrimps, worms and insect larvae.

When it is not feeding the platypus is sleeping in its 7-10 metre burrow which usually has its entrance on the bank of the creek or river, above water level. It can sleep up to 14 hours per 24 hour day.[172] On the rear ankles of both males and females is a spur but only in the male is this venomous; it is one of the few mammals with a poisonous appendage although such a spur is common in reptiles. When the platypus walks it has a reptilian appearance because its body is low to the ground and its legs are at the side rather than underneath. It 'knuckle-walks'—that is, it walks in a manner to protect the webbing between its digits. When disturbed it makes a low, rasping growl.[173] The platypus has a set of molar-type teeth which it loses before leaving the breeding burrow; they are replaced by keratinized pads.

The platypus lays one to three small, leathery eggs, about 11 mm in diameter, which are incubated externally in the nest for about 10 days (they are about 28 days *in utero*). The platypus female is entirely responsible for the feeding and care of the young which is carried out in the nesting chamber of the burrow which she lines with leaves and reeds. The mother curls around the still embryonic young which lap milk which oozes through the skin onto pads on the abdomen. When the female leaves the young in the nest she makes thin mud plugs in the tunnel which connects the nesting chamber to the bank of the stream; she removes these plugs when she returns. Females have 10 X chromosomes and males have 5 X and 5 Y chromosomes (in most mammals the females possess two X chromosomes and males have a single X chromosome and a smaller sex chromosome called Y).[174]

Even with this initial gathering of observations and facts it is possible to begin working carefully and consciously with delicate empiricism and exact sensory imagination. We mentally surround ourselves with the information, a little like in the diagram below with the living animal at the centre. Now we move from one element of information to another, each perceived in the fullest, clearest way possible, in different combinations, around and across the circle—slowly and deliberately proceeding as the inner picture becomes more and more vivid. It is very much as if the different facts and observations are having 'conversations' and seeking to establish meaningful relationships with each other. In our cognitive imagination we need to actually *experience* these interactions as a living process. What is organically the case in the dynamic physical form of the animal we are seeking to recreate in the form of an exact dynamic imagination. In other words, *we are seeking to re-create the platypyus in our cognitive imagination.*

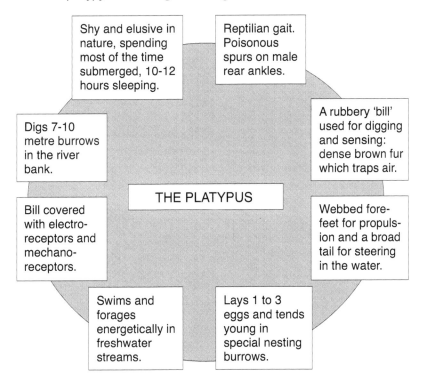

Fig. 45

Relationships such as the following begin to emerge out of our picture: we see a sensitive, shy animal living mostly underwater or underground, the whole body having close physical contact with earth or water almost the whole time. Further, it is active in the darkness or semi-darkness and this, too, acts as a shrouding around its sensitive existence (we shall see in the next chapter that such shrouding, self-enclosing behaviour is typical of the whole 'sense-nerve' group of mammals, in particular the rodents). For a large amount of its time it is asleep which, too, is a kind of darkness. When active in the water, its sense of vision is shrouded by folds of skin and this is when other senses come into play—notably its highly sensitive bill. We see a strong contrast between this intense, sensitive foraging activity in which it eats half its body weight and the deep sleep for digestion, each for half the day. We can say that, in its overall life form, it is a highly sensitive and also powerfully rhythmic animal.

What sort of animal will spend up to four months tending young which, when born from an egg which has little yolk, are at an early embryonic stage, something resembling a slug? As is the case with all marsupials the platypus egg never implants in the uterus, forming a placenta and sharing the mother's blood. The placental animals (like cows) bear their young deep within their unconscious metabolic-reproductive organism so that, when they are born at a relatively much later stage of development, they resemble their parents. With marsupials (like the kangaroo) the embryo is fed with an internal yolk sac; then the tiny embryo emerges from the mother, crawls into the dark protective pouch where it is suckled on nipples for nine months and then is again 'born' when it leaves the pouch. Thus we can say: for the marsupials the womb is without—it is a womb-pouch. Nevertheless it is part of the maternal body of the marsupial and the embryonic form is 'carried' by the mother; it is only *relatively* external. In the next chapter we shall see that the tendency to give birth to young at a very early stage of their development (altriciality) is a characteristic of highly sensitive mammals, including rodents.

The monotreme represents a further emphasis of this 'sensitive' tendency in that the offspring are born as an egg (or, in other words, at an even earlier stage of development than the marsupial). With the platypus there is no maternal pouch-womb as part of the female organism, as with the marsupial—the platypus female must *make* the 'womb' as something external to her body. This she does by greatly extending the burrow after mating and forming the elaborate nesting chamber lined with matted leaves and reeds. The womb of the platypus is the dark enclosure which is the creation of her own instinctive activity; it thus represents the polarity to the mammalian womb or 'true placenta' which lies enclosed, deep in the dark and unconscious process of the female organism.

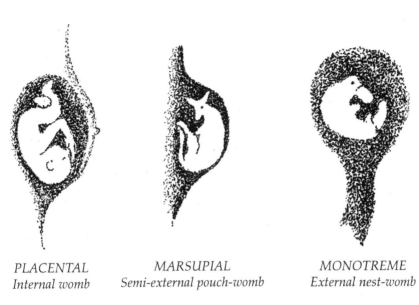

PLACENTAL   MARSUPIAL   MONOTREME
Internal womb   *Semi-external pouch-womb*   *External nest-womb*

Fig. 46

We move carefully through this series (see Fig. 46) with cognitive imagination, forwards and backwards, and experience how the mammalian *typus* creates these three different forms of womb, each as a different expression of the same formative principle. In this way we can comprehend the belonging-together of these three womb-forms in a meaningful way which relates to the whole nature of the animals. By contrast, the classificatory mentality disrupts the living imagination of nature in its wholeness—for example, by saying the marsupial and monotreme are divided from each other based on presence or absence of a pouch without insight into how presence or absence of a pouch *relates* them.

In the above ways we approach an understanding of why the platypus lays eggs. The platypus suckles its young in a way which expresses its uniquely sensitive and highly active nature. The gesture of dedicated busyness which is most notable in its waking character, foraging mainly alone, sensitively and incessantly scanning with its bill the river bed for the slightly signs of animal life, we find also in the dedicated creation of the womb-nest. Even more than the marsupial which still 'carries' its embryonic young in a more or less unconscious way, the platypus is sensitively and devotedly 'responsible' for the embryonic progeny from the moment it leaves the egg. It is thus unjustified, even demeaning, to speak of the egg-laying character of the platypus as 'primitive'—it is simply a feature of 'who' this animal is.

The more vivid our inner picture of the platypus's form and behaviour, the more it becomes possible to imaginatively perceive correlations between different aspects of the animal in terms of the 'law of compensation of parts'. There are a number of features of this animal (apart from egg-laying) which initially seem unusual or even contradictory when it is compared to other mammals. One such feature is the spur on the rear feet of the males which delivers an extremely potent venom; this represents a very strong preponderance, an enhancement of a very specific kind. In our whole-picture of this animal we can discern correlated diminishments, also of a nature very specific to the platypus. Firstly, this animal is highly diminished in relation to its *mobility* on land (where it is most in danger)—it 'knuckle-walks,' body low to the ground and legs to the side, almost like a waddle. It is a very slow mover, with no other bodily defence except its low, rasping growl. Further, in terms of aggressiveness, this animal also displays a great de-emphasis; it exhibits a predominately shy and docile disposition. In the 'budget' of its formative forces, the extent to which this animal is vulnerable and docile is the extent to which it requires an in-reserve weapon of offence (the fact that the spur appears only on the male, which is dominant between the sexes and the 'guardian' of the territory, is inversely correlated with the uniquely sensitive and devoted character of the female).

We find another example of the compensation of parts in relation to the so-called 'duck bill' of the platypus. Even the most superficial observation and comparison to other mammals will conclude that the large bill of this animal represents an extreme preponderance. It is greatly emphasized in being the primary tool for foraging underwater and for digging—the webbed feet have other functions. Further, the bill represents a strong enhancement in terms of the senses; its broad surface is coated with a very large number of sensors which allows a three-dimensional 'fix' on the prey of tiny crustaceans while foraging under water. The bill is an extraordinarily sensitive organ for detecting fine fluctuations of water pressure and electrical impulse connected with this tiny prey. In this regard it can be compared to the highly sensitive nose of the star-nosed mole.[175]

Like the mole, too, the other senses are correspondingly de-emphasized through the principle of 'giving and taking', in terms of both form and behaviour. To the degree that the bill is emphasized in the platypus, the other senses are diminished. The animal dives with its ears closed and its small eyes covered with a flap of skin. It 'sees' primarily with its bill, for most of its waking time. The venomous spur of the male is also inversely correlated to this diminishment of the senses—for this diminishment of a peripheral, broad surveillance by means of the eyes and ears while under water is compensated for by the potent weapon of strike which the spur represents.

When the platypus was first viewed by naturalists—and even still by those reporting on its genome analysis—it was perceived as a 'weird' concatenation of mammal, bird and reptile elements. With our study of the platypus we have a wonderful opportunity to go beyond such tendencies of thinking and cross the threshold to the sphere of its living form. Through exact observational work with the platypus, if carried out with genuinely fresh eyes and an open mind, the being of this animal can unfold like a work of art in the process of creation. It does not become more defined in terms of rigid demarcations or categories but, rather, ever more luminous as the depth of its nature is realized. Then certain prejudices may fall away and we may see the animal for what it is in itself.

For the truth is that only a superficial impression could lead to the idea that it is 'part bird'. Its 'frontal appendage' is certainly not a beak and webbed feet are characteristic of a number of other mammals, including beavers and muskrats. Further, the fact that it lays eggs gives no reason to think it is 'part reptile'—for reptiles are not a defined group at all but only a name that groups unrelated animals.[176] Eggs, cloaca and scales are also features of birds and fish. There are only a few monotreme species but if there were thousands then a platypus egg would simply be a mammalian egg and the spur a mammalian spur. If we were to employ an analysis of sorts using Goethean language we would say that the platypus shows both specialized and general characteristics; egg laying would be general and the spur would be specialized. The platypus is more closely related to a mammal than it is to any other organism; this is because it is a vertebrate with hair and lactating glands. Like every other animal, it is a whole organism, complete unto itself, unique, and this is not merely a theory—we must learn to actually *see* it. Put in the terms of the true comparative biology of which Goethe was a founder, the platypus is a *taxon*—that is, an organism which is whole unto and of itself, but also belonging to a larger *taxon*—the marsupials—which is discovered through empiricism or actual observation of relationships (that is, the primary shared homologues of vertebral column, hair and lactating glands). It is only when species constructions are foisted onto it—mental constructions which group organisms in convenient ways so that they can be explained in terms of their origins—that the platypus ends up out on a limb as a product of parallel evolution or as a mere 'transitional species'.[177]

The classificatory scientific mentality, heightened today by the ease of data collection and manipulation on computers, can stand as an almost impenetrable barrier in the way of a clear seeing of 'who' an animal is. The importance of observational work carried out in the spirit of reverence, the value of a delicate empiricism, need to be understood. In fact all data, all observations, can be of value in the whole-seeing of an animal; everything depends on the care and respectfulness with which the scientific mind artistically builds its exact imaginations.

Chapter 7
Type, Archetype & Archetypal Phenomenon

Intellectual Cognition (Philosophy)

In the literature on Goethe's way of science, references to 'archetypal phenomena' or 'primal forms' are quickly encountered. The notions of type and archetype were central to Goethe's thinking and in this respect he was connected to a very ancient philosophical tradition concerned with how related types of natural phenomena come into being. This was by no means a narrow esoteric preoccupation on the part of Goethe; the idea of archetypes flowed through nineteenth century biological science with its interest in the fundamental patterns or structures apparent in the different groupings (or types) of living things. We need to elucidate Goethe's approach in relation to more general philosophical views on archetypes but always with consideration for what makes the Goethean approach unique—namely, the way it puts these ideas into *practice* in the science of living form. In practice it is impossible to separate the notions of type and archetype from Goethe's thinking on polarity, intensification and metamorphosis.

The idea of the archetype is so closely connected with the philosophy of Plato that any evidence of archetypal thinking in the work of an artist or scientist is usually labelled a 'return to Platonism'. Or otherwise, if one claims to be able to think *in* or *as* the eternal archetypes—as both Goethe and Schelling do—then one may be criticized for assuming too lofty an idealistic position.[178] Kant, before the time of Goethe, had hinted at an *intellectus archetypus* in connection with his discussion of what it means to think the nature of an organism; this he conceived as an intelligence which thinks creatively, from an intuition of the whole into the physical parts—a kind of 'divine reason' as Goethe described it.[179] In effect, Kant was pointing the way towards a true organic thinking (a cognitive feeling and cognitive will) while at the same time denying that this kind of thinking is a human possibility. Kant proclaimed that science can only proceed *as if* organisms are the products of an ideal plan and that the actual task is to analyze living things logically and mechanistically, to seek out the mechanisms of nature.

What Kant did do was to set the stage for an intense consideration during the nineteenth century of the difference between inorganic and organic science. We have already seen in Chapter 2 that Goethe's notion of the *typus* was central to this consideration. The *typus* belongs to the realm of life, not to some separate, metaphysical realm; it is nothing other than the lawfulness of organic form that Kant had pointed to. As discussed in that chapter, the *typus* which 'flows into' the individual living being becomes what we call the identity of the organism, as revealed through its meaningful structure. The archetypal as expressed in the inorganic realm is the subject of the study of physics—here we call it a natural law. A natural law cannot be identified with any particular being; it governs phenomena by standing over them. An inorganic phenomenon *obeys* a law but is not in itself a self-structuring lawfulness (which is how we can define an organism). The view of Goethe and Schelling, and the biologists who followed after them, was that life science *must* proceed to think in terms of the *typus* (not mechanistic natural laws) because only this makes it an authentic science of living form.

It is perhaps difficult to accommodate the possibility that an archetype is not just a philosophical-scientific concept or lofty ideal but an actual cognitive experience—for this is certainly the position of Goethe. He wrote: 'Impelled from the start by an inner need, I had striven unconsciously and incessantly toward primal image and prototype. . '.[180] It is the usual view that the 'divine archetypes' of Plato, like any divine being, are a matter of belief or conjecture. In Goethe's way of science seeing is everything—but we have discussed in previous chapters that Goethe pointed the way to different forms of seeing in scientific practice, to the awakening of different 'organs' of perception, beyond the physical eyes or ears. From the philosophical point of view one may connect Goethe's and Schelling's views on the immanence of the absolute in nature to Spinoza and others. But the main challenge with respect to Goethean science is to cultivate the practice that may open the way to an archetypal thinking. Through working with Goethean science it becomes evident that the archetype is not remote and super-worldly but is part of our immediate, living experience of the world.

Traditional philosophy has produced the conception of a pure mental (divine) model or prototype, a universal ideal, from which physical things are creatively patterned. However, for the logical-deductive approach of conventional science, such conceptions amount to untestable speculation. In relation to Goethe's way of science the philosopher Henri Bortoft explains:

> In a moment of intuitive perception, the universal is seen within the particular, so that the particular instance is seen as a living instance of the universal. What is merely particular in one perspective is simultaneously universal in another way of seeing. In other words, the particular becomes symbolic of the universal.[181]

This makes clearer how Goethe experienced the archetype in his scientific work—as something concrete and in a sense 'visible' to an imaginative or intuitive form of thinking. It has a connection to what the philosopher

Hegel later called the 'concrete universal', the revelation of the universal in the physical particulars. In the experiential or phenomenological approach which awakens the capacity of cognitive feeling, Goethe drew from the philosopher Herder's notion of the *Schwerpunkt* ('centre of gravity') which he had worked out mainly in relation to cultural identity.[182] Herder wrote that by 'feeling into' (*Einfühlung*) the different empirical characteristics of a people we can come to grasp their essential unifying character.[183] For Goethe this helped open the way to a broader scientific practice involving the forms of nature and the human being.

In the first place the senses perceive the form of a mineral, a plant or animal in an exact, meticulous way if such observations are to be part of a scientific experiment. We are able to be exact about such observations because these forms are definite and fixed; the mineral, plant and animal are seen as objective *creations* of nature. The active imagination which then dwells feelingly in these forms begins to grasp the inner creative gesture of the form through entering into the dynamic of polarity, intensification and metamorphosis. When it reaches the stage of cognitive will the subject-object relationship with the form is overcome; then thinking experiences the dynamic or *creative* relationship of the universal and the particular, how the universal is *imparting* itself as the particular part or form. The important point here is the word 'experience'; this is not a logico-scientific deduction from empirical evidence but an actual *seeing*— with the eye of the imagination.

When we look out over the multiplicity of forms of nature and the human world we cannot help but be struck by the abundance of nature's creativity—but to everyday perception the rhyme or reason for this creative abundance is obscure. Within this abundance a phenomenon may, in a sense, come forward and 'speak' directly of a lawfulness which otherwise lies hidden; this is what Goethe called an archetypal phenomenon (*Urphänomen*). Goethe considered such a phenomenon as 'an instance worth a thousand, bearing all within itself'.[184] A well-known example is when Galileo discerned the laws of motion from swinging lamps in a church; this observation shed light on a whole sphere of related dynamic phenomena which is why Goethe called such significant observations 'pregnant points' according to his view that 'nature has no secret which she does not somewhere place openly before the eye of the attentive observer'.[185] Science progresses from such moments of insightful experience; they are the moments, filled with meaning, when the archetype becomes visible to imaginative-intuitive perception and a *'higher experience* within experience' becomes possible.[186]

The magnet reveals polarity in an immediately experienceable way; in or through that physical expression we may grasp the nature of polarity as it works throughout the world—for example in the polarity of Sun and Earth, flower and root, as explored in Chapter 5. The blue colour of the sky reveals in a primal way the relationship of light and darkness (the light-filled atmosphere and the darkness of outer space) and in this experience—without recourse to theory or explanation—we may grasp the nature of colour as such. Rudolf Steiner writes:

> When he encountered a phenomenon, [Goethe] looked for all similar and related facts belonging to the same sphere, so that he would have a complete whole before him. Within that circle, there had to be a principle through which the inner necessity of everything regular—indeed, that of the whole circle of related phenomena—became apparent. To him, it seemed unnatural to explain manifestations *within this* circle by dragging in conditions that lay beyond it.[187]

Here we see the intent of Goethe's science set forth in a perfectly clear way. It shows us that, while philosophy may clarify issues and clear the pathway forward, philosophical discursion is not itself the way of this science. Goethe's science begins in the sphere of phenomena and stays within that sphere; it does not call upon philosophy, any more than it calls upon mathematics or theology, to explicate a phenomenon which is 'speaking' to the researcher in and of itself. The explicating principle, the archetype, is discovered through the experience of the phenomenon—that is, when in this immediate, concrete experience a 'higher experience' becomes possible through a thinking which has transformed from cognitive feeling to cognitive will.

Cognitive Feeling (Goethean Science)

a) Archetype

The task of Goethean science teaching is to bring it about that students gain a rich understanding of the archetype as cognitive experience—that is, as more than just theory. What is set forth below is really an extension of the examples presented in Chapter 2 where Goethe's understanding of the *typus* was first introduced. Any teacher working with the notion of the archetype would need to be cognisant of how long and diverse the tradition of the archetype is in Western civilization and, for this reason, how entrenched certain conceptions can be. In philosophy the archetype tends to be considered as metaphysical, merely 'an idea'; in Darwinian biology it is taken to be the ancestral organism. The artistic approach to this study has the virtue of speaking immediately to the students' feeling understanding.

Musical experience can reveal dynamic archetypal structure in a directly audible way. Examples can be chosen which can be understood whether the student is musically trained or not. The musicologist

Viktor Zuckerkandl relates music formation to the archetypal phenomenon in the Goethean sense; he perceives that every musical work grows out of a seed which he calls the 'primal form' and which is the fundamental law governing the organization of the pattern.[188] Let us consider the classic 'theme and variations', a form used by composers like Mozart and Brahms. Students should endeavour to 'feel into' the music they hear, to find their way toward the 'centre of gravity' or primal form of the piece. To do this they can be played several variations (elaborations) and only then the originating theme; this going backward helps them to discover the archetypal form themselves, perceived (heard) shining through the different variations. Other pieces could be played, such as the *Nocturne in F minor* by Chopin, where the originally stated archetypal melody immediately unfolds in the form of a variation in which the archetype is still present formatively.

Fig. 47: Excerpt from Nocturne in F minor *by Chopin.*

From such musical experiences of the formative action of the archetype the world of plants can be turned to; plant forms also reveal an unfolding in stages (metamorphosis). Like the musical examples, the formative pattern is more or less 'hidden' in the immediate form of the plant. The trajectory of this study beyond merely intellectual understanding needs to be clear; in one way or another the important thing now is to guide the students' thinking into and through the progressions with feeling-thinking, backwards and forwards, as with the progressions of colours and leaf metamorphosis presented in Chapter 5. Any example of a flowering plant can be considered. We carefully observe the form of the plant, from its roots, through its stem and leaf forms, to its floral form; inclusion of the fruit and seeds is preferable. This is the starting point—the plant as physically perceived through a delicate empirical study.

Goethe identified six expansion and contractions of the flowering plant, between seed and fruit. The seed expands into the vegetative organs, contracts in the flower bud, expands into the floral form while it is contracted into the inner floral organs, expands into the fruit while contracting into the new seed forms. These are not physical but formative movements; when they are inwardly, imaginatively enacted a dynamic picture begins to emerge of the plant as a unified, dynamic wholeness.

What the eye of the imagination is observing is a formative theme or archetypal pattern which can be followed into its variations in any number of different species.

The same progression in the flowering plant can be worked with in a different way by focussing on the geometry of the unfolding plant organs—from point (the seed), to line (the stems and petioles), to plane the (leaves), to three-dimensional form (the flower and fruit). This can be experienced in a cyclic fashion, for the fruit produces point-like seed forms. With the eye of the imagination the archetypal theme of plant growth is perceived between the poles of Earth and Sun, uniting them creatively. No matter which plant is observed in the entire life sphere of the Earth, we can 'see' in or through the particulars of that plant a universal drama of growth and unfoldment.

Our study so far has been a way of deepening our vision of the plant archetype. The progression of expansions and contractions, the transforming plant geometry—together these help to cultivate an 'archetypal thinking'. Gradually the physical eye which perceives the processes of metamorphosis in a plant becomes an understanding eye which perceives a universal meaning. The plant is transforming and *intensifying* in its growth towards the flower. The flower as the zenith of the growth process draws both the expansive and contractive elements of the plant into a 'high unity' which we could call more poetically 'a marriage of Earth and Sun'. To enhance our understanding of plant intensification we may now speak of the creative plant archetype in terms of two principles—the axial or *vertical* tendency and the *peripheral* tendency. The peripheral tendency creates the planar leaf and petal forms and the axial tendency creates the seed, stem and inner floral parts.

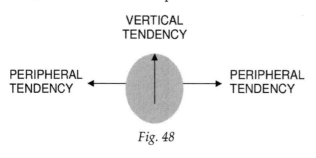

VERTICAL
TENDENCY

PERIPHERAL
TENDENCY

PERIPHERAL
TENDENCY

Fig. 48

The *vertical* tendency carries the plant aloft. As the linear central stem or stems grow they maintain the seed-potential at their tips, the 'growing point' or meristem tissue.

The potential which was in the seed is carried out of the earth and lifted towards the Sun; the growing points of stems are 'elevated seeds'.

In the *peripheral* expanding tendency expressed in the leaves, the creative potential or the idea of the plant is actualized and 'used up' to create the substantial leaf forms. These leaf forms are final in shape and contain no creative potential other than the assimilation of light and formation of food substances. Their only next step is death and falling from the plant which, for the deciduous plants, is a yearly occurrence. Leaves are maximum substance and minimum creative potential.

Fig. 49

When the plant reaches the flowering stage, we see these two tendencies coalesce to create the three dimensional form of the flower. Here the peripheral parts (the petals) encircle and enclose the vertical parts forming the inner organs of the floral cup. In the flower the *peripheral* tendency reaches its zenith in the radiant colour and planar form of the petals. The ability to assimilate substance found in the leaves is lost entirely in the petals—they are thin, insubstantial and, in a sense, close to death (the petals are ephemeral, appearing and soon passing away). Here the archetypal form of the plant achieves its highest stage of actualization, expressed in minimal substance but in greatest radiance, purity and beauty. Gathered around the floral axis to form an often perfumed three-dimensional chalice, the petals are the means by which the previously unrevealed idea of the plant presents itself fully to the external world of animals and human beings.

The *vertical* tendency shapes the organs of the floral axis—the ovary (with its ovules or eggs), stigma, style and stamens (with their linear filaments and anthers which contain point-like pollen grains). The stamen is a metamorphosed petal as Goethe observed in his essay *Metamorphosis of Plants* (1790). In other words, the stamen is a peripheral organ drawn into the sphere of the vertical tendency. The vertical tendency reaches its zenith in the

minute ovules (female seed) and pollen grains containing the male seed;
the original seed-potential, raised up on the stem as the growing point or
meristem, is transformed into the sexual reproductive potential of the plant.
The pollen and eggs are of minimal substance and maximum creative potential which is finally actualized in the fertilization of the egg by the sperm
cells. In this moment the substance nature of the plant is overcome ('matter is spiritualized' to use Goethe's expression); the formative potential (the
idea) is consummated and a new organism is created. The expanding three-
dimensional sustance of the fruit carries this new life in the form of the seeds
which are being prepared to enter the Earth again as seed-potential.

Fig. 50

We have been practising a way of studying plants whereby 'the particular instance is seen as a living instance of the universal'. In plant
growth what is universal or archetypal is the 'theme' of unfoldment
which belongs to plant nature as a whole. With our experience of the
two archetypal creative gestures of peripheral and vertical, it becomes
possible to look out onto the abundance of the plant world and see each
species as some manner of variation on that theme. All plants show a
tendency in one way or the other—either an extreme in either direction
or a balance of the two. Plants which sprawl over the ground—like the
wandering jew—have a minimal vertical tendency and a much greater
peripheral tendency; those whose central stem or trunk soars—like the

tallest of the eucalypts, for example the blue gum—with a less-empha-sized vegetation, have a predominating vertical tendency. In some the expanded quality of the leaf forms predominates while in others it is the flowers.

b) *Typus and type shift*

We work in the same way to understand the unity in the variety of animal form.

In Chapter 2, under the heading of *The Typus*, we explored the penta-dactyl limb form, looking specifically at the limbs of the mole, horse and bat, representing the activities of digging, running and flying respectively. In each case, we experience through cognitive feeling and cognitive will how the pentadactyl form is adapted in a very specific way for a specific earthly environment. With our plastic, sculptural imagination we 'run through' the different pentadactyl forms, re-creating them by imagina-tively sculpting them—compacting them to form the mole paw, empha-sizing one digit and de-emphasizing the others to form the horse's foot, stretching the digits to form the bat's wing. Through this work we come to recognize that *each animal limb is a onesided expression of the same form*—a theme with variations.

It is possible now to enter more deeply into the archetypal formation of the pentadactyl limb in relation to our studies of musical themes and the formative patterning of plants. When any one of the limbs is studied with exact sensory imagination it can be perceived that *the limb as a whole is transformed*—each bone in concert with every other. Entering into the bones of the mole's limb with our exact imagination the same shortened, thick-ened quality is found in each; moving now to the bat, each bone expresses, by contrast, a narrowed, extended quality. Imaginatively moving from one animal limb to the other it is necessary to transform the quality and form of every bone simulta-neously, in concert.

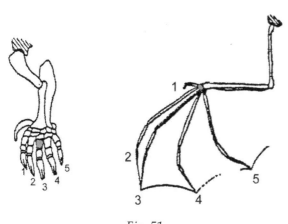

Fig. 51

This perception of the dynamic unity of the parts (which Goethe called the 'correlation of parts') is something entirely different from the 'side by side' logi-cal approach which thinks the organ-ism as being made up of many parts in functional relationships. Through cognitive imagination the unity of the parts of the limb is experienced in just the same way as we experience the wholeness of a work of art, each part relating to every other by an 'inner necessity'. The whole skeleton of these animals could be taken into account in the same way: the pentadactyl system of parts is organized and reorganized in a synchronized way. This has been called a 'type shift'; here we see how the archetype governs the totality of parts.[189]

A small section of a string quartet minuetto may be listened to (Haydn's *String Quartet, Op. 74 No. 1*):

Fig. 52

At the harmonic shift from the Minuetto to the Trio in this section, the instruments change in concert, in a synchronized way, determined by a new mood and a new key. Each of the four 'voices' is more or less unique in its melodic form (most different in the case of the soprano part) yet each is governed and related by the unity of the composition; there is firstly the unity of the Minuetto, then a new unifying impulse in the Trio. What is heard in the music, consciously expressed, is what is at work formatively in the metamorphic processes of organic nature. In the music the type shift becomes an audible phenomenon.

c) Archetypal phenomenon

Artistic study prepares the eye of the imagination for recognizing the archetypal phenomena of nature because it cultivates cognitive feeling in a particularly intense way. Works of art—and especially great works of art—are just that: 'instances worth a thousand, bearing a whole world within themselves', cultural 'pregnant points' which bring to visible expression underlying forces at work in a more hidden way within the cultural period to which they belong. The monumental sculptural work of Pharaoh Chephren (Dynasty 4, c. 2575 BCE) can be considered in this light.

Fig. 53: Chephren on throne with wings of falcon god Horus wrapped around his head, diorite statue, Old Kingdom, Dynasty 4, Egyptian Museum, Cairo.

We work phenomenologically, through exact sensory imagination which develops from a delicate empiricism—an open, receptive way of observing the total form of these works which allows meanings to gradually emerge. We observe the very upright and rather rigid posture and expression of calm repose with the eyes gazing into infinity, the way which the hand is held, and so on. The god Horus is symbolized by the falcon and has its wings wrapped protectively and guidingly around the pharaoh's head, and we know that the head is the seat of human consciousness. Something is being transmitted into the mind of the pharaoh as an inspiration—it is the Word, the divine creative thoughts of the deity which can work productively in the way the ruler can govern the society around him.

The deeper we come to know and understand this sculpture the clearer it shines forth as a representation of hidden forces which are at work in the whole of theocratic ancient Egyptian culture, dispositions and gestures of the inner life of those people in relation to their earthly and spiritual lives. We see here a human being who is not an independent individuality in the way we experience the human being today, who is not possessed of his own sense of self but, rather, is absorbed in the infinite world of the gods. The pharaoh figure, as the key human representative of the time, reveals much about the archetypal soul-spiritual nature of the human being within that culture-whole.

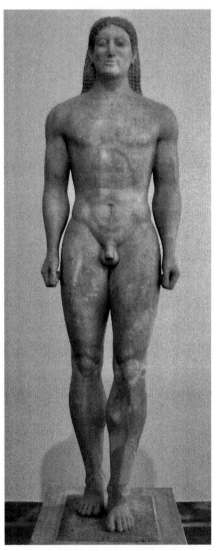

Fig. 54: The Kroisos Kouros from Anavyssos, marble, c. 530 BC. Now situated in the National Archaeological Museum of Athens.

Now some representative works from a later civilization can be looked at. The sculpted figures of Greek civilization express a very different sense of self. In the *kouros* figures of the early Greek period the feet—indeed the whole body—are strong and buoyant; the toes, knuckles, knees and facial features are modelled with a delicate beauty and the chest swells with the breath of an overflowing inner life. This is a human being more centred in himself and expressing an inward satisfaction which comes to the surface in the so-called 'archaic smile'. The human being is no longer entirely absorbed in the world of the gods.

This freedom based on inner bodily experience is far more powerfully experienced in the sculptural figures of Classical Greek civilization such as the Discobolus. The satisfaction and joy of the self at home in the physical body is outwardly expressed in more than a radiant smile—the whole physical body with its grace and beauty of form, its ideal proportion and harmony of gesture, radiates an independence of consciousness.

Yet this human being is still 'ideal' and is not a fully autonomous individuality. The human being is not yet 'his own star'.[190] This quality of coming to be at home in oneself, yet still not in the sense of being an autonomous individual—this is the archetypal quality of the human soul in the era of the ancient Greek, a quality at work everywhere in human life within this culture-whole but in a way which is below the surface of life, covert. This quality is what rises to focalized, directly visible expression in these works, making them true pregnant points.

Fig. 55: A Discobolus in the National Roman Museum in Palazzo Massimo alle Terme. Roman copy of Greek original.

Cognitive Will (Artistic Practice)

Sculptural type shift:
An abstract composition in clay can be created to express a particular mood, atmosphere or gesture—for example, solemnity, purity, expansiveness, directedness—composed of a theme of at least six different elements of form which give expression to this mood (for example, shape, line, texture, concavity and convexity, angularity). This can be conceived as the 'type'. Now, another mood, atmosphere or gesture can be deeply contemplated—then a second sculpture can be made in which all six elements are again represented but which are all expressive of this second mood. All the features which constitute the first type are transformed 'in concert' into the form of the second sculpture: a type shift. A third or fourth sculpture could be carried out in the same way.

Archetypal thinking in painting:
A painting is created on a canvas or paper, using any kinds of paints, with some theme in mind—this could be a specific quality of form, a mood or atmosphere. The painting is carried out slowly and deliberately, with a heightened awareness of exactly what is occurring in terms of the composing of the work. At all times the aim is to experience the whole of the work out of which each element, each stroke or colour, is created; this awareness begins with the whole surface of the blank sheet and is maintained as each element is added. Each new element added should be tested against every other part according to a continual 'correlation of parts'. The aim is not so much to produce a work with a particular content but an experience of the way a finished composition is achieved through a formative idea.

Pregnant point:
This exercise can be carried out with sculpture, painting or poetry—it is a task which will be familiar to an artist. A photograph showing a scene in a foreign land—rural or urban—can be studied in detail, taking in the characteristic forms, colours and perhaps atmospheric qualities. Quite a bit of time should be spent absorbing the qualities of this scene with 'fresh eyes'. Now the artistic work derived from this photograph can begin, with the conscious intention that this work will be a pregnant point, a focalized expression of the forces and qualities which are at work within the scene. The work could be abstract or expressionistic. To help in this exercise it could be imagined that, if someone came upon this sculpture, painting or poem in the middle of this environment, they would soon gain insight into that environment as a whole.

Case Study: The Human Being as Archetypal Phenomenon in The Realm of Animals

Background to this Goethean study

This study arises out of the age-old enigma: what is the relation of the human being to the animals—are we an animal or something other? Scientifically speaking—that is, logic-based, mechanistic science—we are certainly classified with the animals and in terms of conventional biology there is no other possibility. Yet human beings intuitively, naturally, experience themselves to be more than just another animal species. This question is fraught today for it often seems that to claim we are something more than an animal is to be suggesting human superiority over the kingdom of animals. Goethean science leads us, not to a theoretical or philosophical solution to this enigma, but to a living understanding of the human being and the different animal species. We come to experience the human being as an archetypal phenomenon or 'pregnant point' in relation to the animal world, 'an instance worth a thousand, bearing all within itself'.

Goethe was delving deeply into the question of animal and human nature in his mid-twenties, a decade before his journey to Italy during which he made his discovery of the archetypal plant. He was seeking insight into the archetypal form of the animal, building on his extensive studies of human anatomy and observations of animal bones, the form of which he sought to read as a kind of 'text'.[191] In his time natural philosophers were intent on finding what separates us from animals in physical, anatomical terms; the prevailing belief was that what distinguishes us is the lack of one small jaw bone—the intermaxillary bone. It was Goethe who was able to prove that humans actually possess this bone, that the uniqueness of the human being must be sought on another level. He saw that what distinguishes us from animals can only be properly understood through a living, imaginative form of thinking, a thinking which is able to perceive *how animal form intensifies into the form of the human being*.

It was through his connection with his colleague Herder that Goethe was able to come to clarity on this issue. Rudolf Steiner summarizes the ideas of Goethe and Herder as follows:

> On the lower levels of animal organization [the typical form or *typus*] always realizes itself in a particular direction, and develops toward it in a very pronounced way. As this typical form rises toward the human being, it gathers all the formative principles developed in a one-sided way in the lower organisms (which it distributed among various beings) into *one* form.[192]

We have already explored the kind of creative thinking necessary to grasp the animal *typus* in Chapter 2. What Steiner means by '*one* form' is the human being as an archetypal phenomenon in relation to the animals, in which all tendencies are harmoniously gathered and expressed. It was an idea, as Steiner goes on to say, which had 'an uncommonly fruitful effect on subsequent German philosophy'. Our purpose in the study presented here is not so much to develop this idea philosophically as to explore— through cognitive feeling—how we come to actually *perceive* the intensification of the *typus*, phenomenologically, through observation of anatomical and behavioural features.

It needs to be added that this understanding of the relationship of the human being to the animal kingdom by no means merely belongs to Goethe, Herder and German philosophy. On the contrary, it brings forward an entirely different picture of evolution to that of the Darwinian hypothesis, one that has immense significance for our modern civilization and its relationship with nature. What becomes evident is that the human being did not evolve *out of* animal life at all, radical though this sounds at first. Rather, every form of animal is a partial expression of the human being—or, better said, the creative *typus* which is fully expressed or realized in the human form has been active throughout the evolutionary process and every animal is a partial expression of it. In a sense we can 'find' the human being if we consider the animal kingdom as a whole. A careful account of this picture of evolution is presented by Jos Verhulst, focussing on the primates, in his *Development Dynamics in Humans and Other Primates*.[193]

Polarity and the typus

We begin with the human being in its broadest outlines and from the perspective of the human '*one* form' we can look outward upon the multifarious animal forms. We look with 'fresh eyes' at the human shape and see three broadly differentiated centres—the rounded head and the linear limbs, connected through the trunk.

The head is the centre of the consciousness aspect, meaning the senses and the nervous system generally with its seat in the human brain. The human being experiences the world, reacts to it, reflects on it in a detached way through its sensory-nervous functions. We will call this the 'sense-nerve pole'. The limbs are how the human being physically inserts itself into the world; this is not an activity of reacting or reflecting but of physical doing or making. We will provisionally call this the 'physical pole'.

These observations are 'simple' but they are not meant to be merely factual; the intention is to begin to build an inner picture of the two living gestures involved here—sensory-mental response to the world and physical insertion in the world. These two polar orientations are worked through with cognitive feeling, moving between them to come to an inner experience of both their polarity and their unity.

Now we can consider a number of animal forms in the same light—for, most certainly, all animals reflect this polarity, in one direction or other 'in a very pronounced way'. Animals express sense-nerve functions similar to or at least related to those of humans—sight, hearing, taste, smell and other senses which are only just beginning to be understood, such as the sense of gravity and pressure. If we take the catfish (order Siluriformes) we see an example of an immensely heightened sense of taste in relation to the water habitat. We could say that it is a 'taste specialist'. A six-inch catfish has tens of thousands of taste buds which cover its entire body, inside and out, including its fins, back, belly and tail—making it literally a 'swimming tongue'. In comparison, humans have around 10,000 taste buds.[194] All fish express such highly specialized organs of taste to a great or lesser extent.

Insects, too, have taste or chemical sensory cells all over their body, on their antennae and even on the lower part of the legs. For example, the moth has phenome chemicals on its specialized antennae which can 'taste' a mate three or more miles away.[195] The annelids or 'segmented worms' have specialized light sensors and chemical sensors on their heads, appendages and various parts of their body; in that respect their body is 'all eye'.[196] One may look at other groups, such as the molluscs and echinoderms, in the same light.

If we take the eagle as the next example, here we have an animal whose senses are highly specialized in relation to its airy environment. The wedge-tailed eagle (*Aquila audax*) is the largest bird of prey in Australia, found also in southern New Guinea. Its eyes can see with eight times more detailed resolution than humans. This means that a wedge-tailed eagle could see the text of a newspaper from over one and a half kilometres away. They can also see more colours than humans can. They use their well-developed eyes to search for prey and to see rising thermal air currents so that they can use these to gain altitude while using very little energy. Their keen eyesight extends into the infrared and ultraviolet bands.[197]

From these examples it would be possible to move through the entire animal kingdom, considering animals belonging to every manner of habitat—watery, airy or terrestrial—and showing how the senses are specialized in relation to the functioning of those animals within their habitat. In every case of the so-called 'lower animals' like molluscs and fish the whole body acts as a sense organ whereas in the 'higher animals' like the mammals, the senses of hearing, sight and taste are focussed in the head yet are nevertheless highly specialized in different directions—for example, the dog whose sense of smell is 1,000 to 100,000,000 more sensitive than a human's (depending on the breed).[198] By contrast, the human senses are relatively restrained, held back and balanced, without one being especially developed in relation to the others.

If we now turn to the 'physical pole', the way animals are inserted into the physical world through their physical activity, we arrive at a similar picture. Many of the 'lower animals' like the molluscs and annelids (but also reptiles such as snakes) do not have differentiated limbs but, rather, use their whole body as a tool of locomotion. It is appropriate to speak of such animals as 'all foot' or 'all limb'. With the majority of land-dwelling vertebrates we find true limbs, each limb-type specialized in relation to a particular terrestrial environment. We have already looked—in Chapter 2 and above—at how the mammalian pentadactyl limb *typus* is 'stretched', or 'refined', 'concentrated' or otherwise emphasized for burrowing, running and flying in the case of the mole, horse and bat respectively.

The human hand (as with the senses) is 'restrained' in comparison with the highly specialized limbs of the animals—or, it could be said, it is 'specialized to be non-specialized'.[199] Indeed, the hand (as with the other parts and organs of the human physical organization) demonstrates paedomorphism; that is, it retains the juvenile limb form into adulthood. Whereas the limbs of animals, developing from the juvenile to adult forms, lose the potential for a range of movements or actions and become 'fixed' in a particular highly specialized direction, the adult human hand does not lose this potential. Writes Verhulst: 'In the course of anthropogenesis, the hand never loses capabilities but always gains new ones.'[200]

We can summarize this section by saying that the formative principles connected with the essential polarity of the *typus*—the different senses representing the 'sense-nerve pole' and the different pentadactyl forms representing the 'physical pole'—are expressed in 'particular directions', in onesided or 'very pronounced ways' in the different animal forms. By contrast, in the human being the senses and the limbs are formed in a relatively simple, undeveloped manner. The reason for this is evident: with the human being no sense organ, no limb, is adapted or specialized 'in the direction' of any specific terrestrial environment.

The human being as revelation of the threefold typus

Now we can deepen our picture of the mammalian *typus*. We have seen how the *typus* is expressed in the human as a polarity which we called above the 'sense-nerve pole' and the 'physical pole' (but in a onesided or exaggerated way in separate animal groups). Following from our study of polarity in Chapter 5 we can now further show that polarity explored above is in fact a threefold—two poles and the mediating element or copula. In the case of the human form the copula is focussed in the mediating form of the trunk in which qualities which belong to each pole are united.

Credit must be given to Rudolf Steiner for bringing forward the picture of the human as a threefold being.[201] The skull is largely rounded in form, shell-like, with non-moving parts (other than the jawbone or

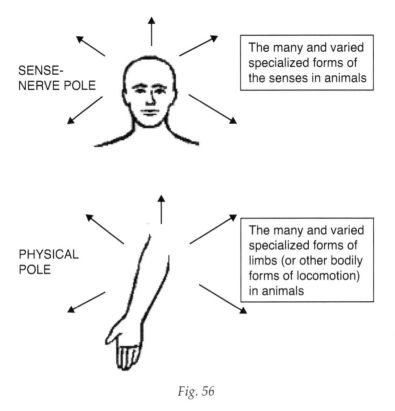

SENSE-
NERVE POLE

The many and varied
specialized forms of
the senses in animals

PHYSICAL
POLE

The many and varied
specialized forms of
limbs (or other bodily
forms of locomotion)
in animals

Fig. 56

mandible). The skull *encloses* the soft tissue of the brain which is the seat of consciousness—the head is the seat of what he calls the *sense-nerve organization* which extends throughout the body. The head is held aloft on the spine in a way which allows it to be relatively motionless and reflectively detached from the external world. The brain is relatively non-regenerative nervous tissue; it is encased in the skull and floating in cerebrospinal fluid, motionless so as to allow *thinking* to take place. Thinking is activity on a higher level than physical activity—concepts, ideas, insights are in themselves non-physical.

The limb bones are mostly linear and elongated with moving joints, embedded *within* the soft tissue of the muscles which are the basis of bodily action. We see that the metabolic organs and limbs work together to allow the human being bodily insertion into the external world through physical activity of every kind. The 'seat' of this *metabolic-limb organization*—as he calls it—is the liver which, through the circulation of the blood throughout the body, enables this physical activity through the production of energy substances. In contrast to the nervous tissue, the liver is highly regenerative and involved with the transformation of substances. This organization is the seat of the power of *willing* (or doing).

Mediating between the above two poles is the *rhythmic organization*. The ribcage is a rounded form, enclosing the soft tissues of the heart and lungs; to this extent it is like the skull. However this form is made up of long, linear bones—the ribs; to this extent it is like the limbs. The spine also has this same mediating, combining character. Each vertebra has a rounded, hollow centre, containing the soft tissue of the spinal cord; to this extent it is like the skull. Each also has linear structures radiating from it; to this extent it reflects the linear limbs. The vertebral column as a whole and the ribcage show a tendency toward a rhythmic structure of repeating units or segments. The overall character of this middle realm is rhythm, predominately expressed in the pulsating activity of the heart and lungs. In the interacting rhythms of the breath and the blood motion combined, there is a mediation between the outer world (through the breath) and the inner world of metabolism (through the blood). Rhythm in this organization is a constant balancing between stasis (sense-nerve character) and movement (metabolic-limb character). The trunk is the seat of the sentient power of *feeling*.

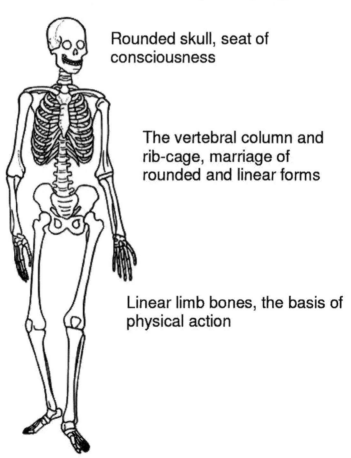

Rounded skull, seat of consciousness

The vertebral column and rib-cage, marriage of rounded and linear forms

Linear limb bones, the basis of physical action

Fig. 57

In the study of Australian mammal forms documented below we are seeking not to merely explain or theorize but to cognitively *experience* that the human being is indeed an archetypal phenomenon in relation to the animals. The threefold archetypal form which is a harmonious 'one' in relation to the human being is in a sense 'spread out' or separated in different animal forms which demonstrate the character of the sense-nerve, rhythmic and metabolic-limb organizations in a onesided or specialized way. Wolfgang Schad has carried out a comprehensive study of mammalian threefoldness in his *Understanding Mammals* and the research presented below is really only an extension of the work he has pioneered. [202] The most important thing, in relation to an orientation course of study, is that research is carried out in a way which unites the three aspects: physical thinking (philosophy), cognitive feeling (Goethean science) and cognitive will (artistic practice). So, for example, it would be vital that students consider examples such as the Greek sculptures as 'pregnant points' as presented above, as well as artistic exercises of various kinds, involving perhaps sculptural, musical or poetical forms, before they embark on such research as that which follows. The study below may have value in terms of an understanding of Australian fauna but, in relation to university orientation studies, its principal purpose is to develop a living, archetypal form of thinking.

A look at three Australian marsupials—the antechinus, thylacine and kangaroo

This research builds on the methods of exact sensory imagination and delicate empiricism which were developed in Chapter 6 in the study of the platypus. Wolfgang Schad has dealt mainly with the European and African placental mammals (in placentals the foetus is brought to term in the womb). He has revealed how, most typically, the rodents (such as a mouse) exhibit the character of sense-nerve animals, the carnivores (such as a leopard) primarily the character of rhythmic animals, and the ungulates (such as a cow) primarily the character of metabolic-limb animals.[203]

The research presented here considers three Australian marsupials (mammals in which the foetal form of the animal is suckled externally in a pouch). A striking fact is that out of the 158 species of marsupial in the world, all except one—the American opossum—are native to Australia and New Guinea, a fact which has led to the view that the Australian fauna is something isolated and even 'strange', as noted in the last chapter in relation to the platypus study. With this preconception in mind we can take the opportunity to look at these animals with fresh eyes, through a living, imaginative form of thinking.

The strong correspondence which exists between the marsupial and placental animal types has been noted by many evolutionary biologists.

There is a placental mouse and a corresponding marsupial 'mouse' (actually called an antechinus); there are also corresponding mole, dog and cat types—even though the placental and marsupial pairs are not genetically related in any known way.[204] It is assumed that the two animals of the pair—say, the mouse and the antechinus—belong to two completely different but parallel evolutionary streams and converge due to natural selection in related environments. The demands of the 'whole animal' approach taken here help us to resist the temptation to rush into evolutionary theorizing and instead to focus on the care and respect required to see an animal in terms of itself. Whether the marsupial mouse and placental mouse developed in parallel or convergence is not crucial at this point; what is important is simply to be able to experience how each animal represents a different way of being a mammal.

As we consider different features of the antechinus (genus *Antechinus*, 11 species) and permeate them with cognitive feeling, the picture grows of a sense-nerve animal of a particular emphasis—in some ways similar, in other ways quite different to a rodent. Taxonomically the antechinus is not a rodent, in spite of the fact that it looks like one and that about 40 percent of mammal species are rodents.[205] The dentition immediately shows this to be the case according to the conventional way rodents are classified—namely, a single feature which is continuously growing incisors in each of the upper and lower jaws (which wear out continuously as the rodent gnaws seeds, nuts, wood etc). The antechinus has no such incisors but, rather, the dentition of a carnivore with four pairs of sharp canine-like teeth with which it pursues beetles, spiders, amphipods and cockroaches.[206] Carnivorous though it may be, the antechinus nevertheless displays the same sense-nerve feature as the rodents which is that these animals must actively, energetically and with heightened senses search for their food. This characteristic is polar to the metabolic-limb nature of animals like the cow which grazes in a slow and languid fashion. As Schad puts it, 'Mice must search for their food but cows find it growing all around them.'[207]

The antechinus has a grey-brown back (dorsal) coat and a paler under or ventral side, with a tail as long as its body ('the development of the tail is directly related to the level of sensory alertness' writes Schad).[208] These features go along with the very diminutive stature of this animal (about 100 mm nose to tail); it moves rapidly and nervously, close to the ground with its dark plain coat allowing it to remain quite obscure (cryptic coloration) as it darts in and out of the bushes on its sensitive, delicate feet. Its nocturnal habit enhances its obscurity. Large protruding eyes, a long pointed snout and long whiskers, large hairless ears, 'are signatures of an overly sensitive nature'.[209] It occupies communal nests in hollows or forks of eucalypt trees; such nests, writes Schad, of animals with a strong

Fig. 58: Brown antechinus (Antechinus stuartii).

sense-nerve orientation, are external sheaths which act as supplementary bodily coverings and allow the animal to withstand 'the onslaught of sense-impressions'.[210]

A notable fact concerning the antechinus is that, at the end of the two-week mating season in spring in which males mate promiscuously, the males all die.[211] This too is a sense-nerve characteristic and relates in general with the tendency of sense-nerve animals (like the rodents) to die easily. Schad notes that mice can die from a severe fright and the mass self-extermination of lemmings is well-known.[212] In all these ways the antechinus is very similar to mice and other rodents.

Now we come to the main feature which supposedly sets off the marsupial antechinus entirely from its placental 'twin' leading to all the postulations about parallel and convergent evolutionary development—this is the fact that it rears its foetal young in an external slit-pouch. Actually this feature, seen from the 'whole animal' point of view which doesn't isolate characters from the whole, is only an exaggeration of a typical sense-nerve characteristic of rodents; namely, that all new-borns bear little resemblance to their adult parents and are more like foetuses (altriciality). 'Mice seem to be in a tremendous hurry to raise their young,' writes Schad.[213] Schad remarks that the sense-nerve animal lacks the metabolic strength to devote itself to the development of its young and that this may be contrasted to metabolic-limb animals like a cow which 'carries her young for an extended period (280 days), shaping it deep in the unconscious reaches of her body until it is almost completely developed—actually already "too finished"—before she finally has to give birth'.[214] Schad goes on to say that the sense-nerve animal replaces the uterus with a nurturing from without, often within nests, 'through her own nervous activity in the outside world'. We can thus conclude that the marsupial nature of the antechinus represents a strong and particularly emphasized sense-nerve characteristic, an extreme form of altriciality (we have seen in Chapter 6 that the egg-laying monotreme goes even further in the direction of this tendency).

Fig. 59

Turning now to an apparently (and recently) extinct marsupial—the so-called Tasmanian tiger or thylacine (*Thylacinus cynocephalus*)—we have before us an animal which, like the other Australian marsupial fauna, has become mired in preconceptions. In the first place it was called a 'tiger' because of its stripes but, patently, it was a dog type and its dentition is in fact very close to that of the red fox (*Vulpes vuples*)—yet, unlike a dog or fox, it had a pouch in which up to four joeys would be nurtured for around three months.[215] Such diverse features have made it appear an oddity, a 'living fossil', developing in isolation as in the 'evolutionary backwater' of the Australian continent since the break up of Gondwanaland millions of years ago—viewed from a northern hemispheric, placental point of view. In this connection we may consider the comments of biologist Richard Dawkins on the thylacine:

> To any dog-lover, the contemplation of this alternative approach to the dog design, this evolutionary traveller along a parallel road separated by 100 million years, this part-familiar yet part utterly alien other-worldly dog, is a moving experience.[216]

In fact the thylacine is not 'alien' or 'separated' in its development at all; it is simply a unique creation of the mammalian *typus*. The thylacine is a particular expression of this *typus* in relation to the Australian landscape—strongly emphasizing the middle or rhythmic aspect yet having a particular tendency toward the sense-nerve character which typifies the Australian marsupial fauna as a whole.

In terms of relative sizes, it was a 'middle' marsupial, standing 60 cm at the shoulder—mid-way between the diminuitive sense-nerve antechinus and the larger, more metabolic-limb kangaroo. The rhythmic or 'middle' character comes through clearly in the carnivorous habit of the thylacine; it hunted for kangaroos, wallabies and other smaller mammals. The thylacine was a nocturnal and crepuscular (dawn or dusk) hunter, spending the daylight hours in small caves or hollow tree trunks in a nest of twigs, bark or fern fronds. It tended to retreat to the hills and forest for shelter during the day and hunted in the open heath at night.[217] The stomach was muscular and capable of considerable distension after taking in a great amount of food. The thylacine was not able to run at high speed but able to leap upon its prey.[218] In these last features it was close to large cats like the lion and leopard—rhythmically mediating between the sense-nerve aspect of the highly watchful and active hunt, then gorging, leading the metabolic-limb aspect of a quiet and lengthy digestion process.[219]

Schad writes that a metabolic-limb animal like the cow 'gives the impression that its life is buried deep within its massive body' and that '[its] visible surface . . . is only the indifferent covering of a rich, autonomous inner life'.[220] The exact opposite is the case with sense-nerve animals like rodents which normally show a sharp division between dorsal and ventral colours; such an animal 'seems to live more outside its body than within it'.[221] In the case of the 'middle' or rhythmic animals—notably the carnivores with their typically striped or spotted coats—the surface of the body speaks of a life which is 'neither so completely preoccupied with its own metabolism as the bovid, nor as entirely devoted to the outer world of the senses as the rodent', comments Schad. 'In these animals, more than any other, the surface of the body speaks directly through the intensity of its dynamic colouration.'[222] Thus we note the conspicuous rear stripes of the thylacine and find our way through cognitive feeling towards understanding this patterning in relation to degree and quality of the rhythmic constitution of the animal.

Fig. 60: One of the last photographs (1928) of the thylacine (Thylacinus cynocephalus) showing the rear lower leg resting on the ground, similar to a kangaroo.

Although the thylacine demonstrates a primarily rhythmic character, this is orientated strongly in the sense-nerve direction. We have already noted its 'shy' and 'secretive' nocturnal habit and the fact that it retreated to more or less hidden nests in the daylight hours—these are sense-nerve features. The sense-nerve aspect is also evident in the fact that this animal had notably large pads on its rear feet and that the lower part of the leg could rest entirely on the ground. The thylacine was noted for its rather awkward gait (making it a slow runner) and that on occasions it could sit upright and hop remarkably like a kangaroo.[223] Striking also is the protrusion of its rump and how this appears to extend into the thick long tail; also the fact that this is the only marsupial with a rear-opening pouch.[224] These features (together with the appearance of the stripes on the rear part), shows a particular emphasis on the posterior pole of the animal which is characteristic of sense-nerve animals like rodents (and polar to the metabolic-limb animals like the cow or ox, where the frontal part is emphasized, in shape and bulk and in structures like horns). Schad notes that, while the *physiological* seat of the sense-nerve organization is in the head, the *physical* emphasis of this type of animal is the posterior pole.[225] The ability of the thylacine to sit on its haunches, head alertly in the air, and to spring energetically, have the qualities of a sense-nerve emphasis which we also find in the kangaroo.

Other characteristics of the thylacine also speak of a sense-nerve emphasis to its primarily rhythmic orientation; its cry was a series of short, cough-like husky barks (compared to the fulsome roar of a lion or the deep 'metabolic' moo or bellow of a cow). We have already observed, in relation to the antechinus, the sense-nerve character of giving birth to foetal young and rearing them externally in a pouch, noting that these animals have not the 'metabolic strength' to raise young internally in the womb. The thylacine reared its cubs in its pouch for about three months.[226] We cannot compare the nature of the thylacine to the extreme sense-nervous character of the antechinus. Nevertheless, as we build an exact imaginative comprehension of this animal by bringing its different aspects together into an inner picture, we can begin to see exactly how its sense-nerve aspect lives together with its primary rhythmic character in the particular emphasis or direction which has to do with its relationship with its native Australian habitat.

Fig. 61: Eastern grey kangaroo (Macropus giganteus).

The last marsupial to be considered here is the eastern grey kangaroo (*Macropus giganteus*). Standing around two metres tall, it is the second largest marsupial in Australia (the red kangaroo being the largest). Its range covers about a fifth of the eastern Australian continent. It is a grazing animal, eating a wide variety of grasses on the dry, open plains where it lives in groups (which are called either herds, troops or mobs). In these features we already see clear signs of the primarily metabolic-limb orientation of this animal; with respect to its size and communal grazing habit it is a relatively 'slow' and languid animal (compared to the constant nervous search for food which characterizes the diminutive antechinus). Further, as with metabolic-limb ungulates like cattle, the kangaroo has a highly specialized digestive tract for the digestion of cellulose; it has two stomach chambers, the sacciform and the tubiform, the first containing an abundance of bacteria, fungi and protozoa that begin the cellulose fermentation process. Food may remain in this part of the stomach for many hours and, like a cow chewing cud, the kangaroo may spit up bits of undigested food to be chewed and then swallowed again.[227] This powerful inwardness of metabolic life we can add to our picture of the metabolic-limb orientation of this animal. The animal crops grass with its incisors and grinds it with molars which are replaced as they wear down; this emphasis on molar formation in the dentition closely resembles metabolic-limb animals like cows. [228]

However, we may go only so far in a comparison of the kangaroo with the ungulate. For, indeed, like all the marsupials, there are numerous other characteristics which confer on these animals a strongly sense-nerve orientation. The eastern grey kangaroo is nocturnal and crepuscular, coming onto the plains to feed especially in the early morning, and during the daylight hours retiring 'shyly' to wooded areas. When disturbed it sits up alertly on the extended lower part of its rear limbs with its large, sensitive ears held erect and can hop extremely rapidly with the help of its muscular tail. In terms of body mass and overall shape, the physical pole of the animal is posterior and this we have seen is a characteristic of sense-nerve animals. It has some suggestion of a 'sense-nerve' white coloration on the ventral part of the body. The kangaroos, like all marsupials, give birth to young in virtually embryonic form; this young then finds its way to the external pouch where it is suckled. We have already seen that such altriciality represents a strongly sense-nerve characteristic; the life of the sense-nerve animal is developed more externally than metabolic-limb animals like cows with their deeply inward, unconscious forces of growth and sustenance.

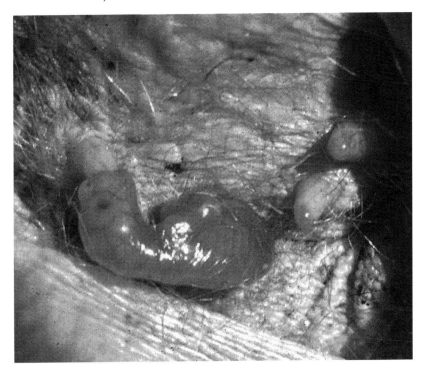

Fig. 62: The very young joey of the Eastern grey kangaroo in the pouch.

Thus the eastern grey kangaroo represents a unique emphasis of metabolic-limb and sense-nerve aspects particular to its life in the grassy plains and wooded country of eastern Australia. The mammalian *typus* has been drawn formatively in this direction by the specific qualities of this landscape—which is another way of saying that the particular metabolic-sense-nerve character of the kangaroo is an expression of its habitat. The animal *is* its native habitat in a focalized form. We are thus able to speak of the coming-into-being of this animal without recourse to an evolutionary theory pressed, as it were, onto the animal from without. What we can learn about the kangaroo—as with every form of animal— comes from within the life-sphere of the animal itself. Through a cognitive feeling becoming a cognitive will, which participates imaginatively in the form of the animal, we can begin to comprehend the creative activity of the mammalian *typus*.

We may now draw this exploration of Australian marsupials to a conclusion in terms of our theme which is the human being as archetypal phenomenon in the realm of animals. With the eye of the imagination we look out into the Australian landscape through its animals—here through the form of the antechinus, thylacine and kangaroo—and in a sense which belongs only to this imaginative outlook—we are at the same time seeing ourselves. We are seeing three expressions of the animal *typus*—gathered in a unity in the human form and spread out creatively in these particular forms. These animals are manifestations of the sense-nerve, rhythmic and metabolic-limb aspects, with *each* nevertheless revealing an orientation towards the sense-nerve aspect. Returning once again to the words of Steiner, in these animals we see the *typus* 'realize itself in a particular direction, and develop towards it in a very pronounced way'.

PART III
THE ECONOMIC SPHERE: A GOETHEAN TEACHING METHODOLOGY

'We can never found a real science of economics without developing pictorial ideas; we must be able to conceive all details of our Economic Science in imaginative pictures. And these pictures must contain a dynamic quality . . .'

Rudolf Steiner[229]

'Citizens, no matter what happens today, in defeat no less than victory, we shall be making a revolution . . . it is the revolution of Truth. In terms of policy, there is only one principle, the sovereignty of man over himself, and this sovereignty of me over me is called Liberty. . . The common law is nothing but the protection of all men based on the rights of each, and the equivalent sacrifice that all men make is called Equality. The protection of all men by every man is Fraternity, and the point at which all these sovereignties intersect is called Society.'

Victor Hugo[230]

'Everything in economics is invisible. Prices, Values, Capital—none of these things has sense-perceptible existence. And because the things of economics are invisible, the content of economics has to be grasped through the imagination.'

Christopher Houghton Budd[231]

'. . . a spiritual contemplation of nature will provide means for the kind of training in thought which, among other things, makes it possible to comprehend the social organism'.

Rudolf Steiner[232]

Chapter 8
The Metamorphosis of Capital and The Meaning of Money

Introduction

The research documented in this chapter is an examination of fundamental economic phenomena using imaginative cognition and developed through a Goethean methodology. As such, it could become part of a preliminary teaching in economics at university level. As stated in Chapter 4, the form of this research locates it between the orientation course and the teaching of economics as a specialized tertiary course of study within the organic university—in Figure 15 its approximate location is indicated by the grey spot. It is an extension of the life science approach taken in the orientation course and for this reason it follows the same form as already presented in Part II, in relation to the thinking of the whole human being: philosophy (intellectual cognition), Goethean science (cognitive feeling) and artistic practice (the cognitive will).

The science of economics, like the biological sciences, emerged out of philosophy during the eighteenth century—economics out of moral philosophy and the biological sciences out of natural philosophy. The early political economics of Adam Smith was worked out in specifically moral terms, where capitalism is seen as an ethical project to be realized through a political commitment to justice and freedom; the biological sciences had their roots in classical epistemology and metaphysics. Both economics and biology set about achieving the status of a science by adapting themselves to the hypothetico-deductive methodologies which became refined into powerful tools in the aftermath of the Enlightenment. These were the methods of mathematical physics—which is to say, of inorganic science. Since that time economics, like the biological sciences, has evolved into sophisticated disciplines through the application of cause and effect logic and the uncovering of mechanisms in order to make all phenomena explicable and predictable. Thus, in modern economics for example, we have the 'market mechanism' and the 'price mechanism'; in biological science 'the mechanism of natural selection'. The challenge today and for the future in relation to tertiary education is to find our way towards a study of economics through the methodology of a true life science.

Intellectual Cognition (Philosophy)

Economics is but one dimension of the social reality. No more than the limb of an animal can be meaningfully studied when isolated from the whole living animal can the economy be comprehended in isolation from its social context. Thus, the first thing to be explored is the nature of society as a whole, for this will determine the way we go about studying its elements or aspects. As discussed in Chapter 2, the method is determined by the character of the phenomenon being studied. Is society a living being and, if so, in what *sense* is it living? Is it possible to speak meaningfully of the living wholeness of society as we can, for example, speak of a living whole like an animal? We commonly use expressions such as 'body politic' or 'social organism' giving expression to the instinctive sense that society is a form of living being; yet it is by no means clear how the methods of life science could be applied to society overall and specifically to economics.

The conception of society as an organism has a long intellectual history in Western civilization.[233] It can be traced back at least to Plato who, in his *Republic*, compares the divisions or estates of society with the three parts of the human soul (the thinking, feeling and willing) which, in ideal circumstances are harmoniously related, each part doing its job well for the common good. Plato's notion of society-as-organism inspired many developments in social and political philosophy down the centuries and had an influence on the struggle of the Romantic thinkers against the mechanical world conception promulgated by the Enlightenment. These thinkers rebelled against the Enlightenment concept of the 'state machine' which viewed the function of the paternal, absolutist state as merely a way of providing security and satisfying individuals' material needs.[234] Wilhelm von Humboldt notably, in his *The Limits of State Action*, advocated the minimalization of state power in order to allow for the possibility of individual self-cultivation.[235]

The question of freedom was a pre-eminent concern of the Romantic thinkers in their considerations of the social organism, even though 'freedom' seems to contradict the idea of 'society as an organism'. For, in any biological organic structure (like the cells or organs of an animal body) each part serves every other cooperatively, without the least volition or freedom in relation to its activity. The meaning of organic interdependence of parts appears to depend on a *necessity*. Hegel, in his *Philosophy of Right*, built his conception of the properly working state as an organic entity precisely on the idea of how individual freedom can be reconciled with the freedom to be a social member and to be at home in society. Hegel rejects and regards as impoverished the 'negative' view of freedom as being the extent to which we, as isolated individuals, are *not* interfered with by external coercion (freedom *from*); the positive view is that individual freedom develops in a healthy way into a freedom-*for*-others.[236] He is, in other words, speaking of the freedom which brings about cooperation and social coherence.

Outside philosophy as such, in the realm of sociology, psychology and biology, the relationship of society and organism tends to be more one of analogy. The idea of the social organism had a notable exponent in the nineteenth-century sociologist Herbert Spencer who, in his *Principles of Sociology*, coined the term 'super-organism' to refer to the interdependence of individuals and groups in society as similar to organs within a living body. A concomitant of this idea is that superior and lower class divisions are 'natural' structures.[237] Spencer did not say that the entity we call a society *is* a living being; rather, that it is *like* an organism. His idea of the super-organism was taken over by biology to describe natural entities such as termite nests and bee hives, where the colony acts as an organism—that is, displays coordinated interdependence of its parts or organs—despite each animal's physical individuality. The idea of human society as a form of super-organism has also received support from areas of human psychology concerned with religion and social identity. The difficulty of connecting human societies with animal colonies remains that of freedom; for example, people (unlike animals) can identify with different groups and break their attachment to them at will.[238] This leads us back to Hegel's view that ideal human social structures (the organic form of the social order) can be *created* out of free action (freedom *for*).

What we come to through such considerations is that human society is a super-organism *of a certain kind*, quite distinct from those which belong to the animal world. The formation which we call human society is through and through *human* in character and reflects what is unique in the human body-soul-spiritual constitution. Society is made up of a multitude of living human beings but it is more than this, because humans are not organisms like plants and animals—they are soul-spiritual beings who say 'I', who possess consciousness of self. The form of the animal super-organism relates to instinctive functions or roles—for example, the queen bee, the worker bees and the drones. By contrast, the overall form of society has what we could call a 'soul constitution' meaning that it is shaped as the expression of the fundamental human driving forces or ideals. As discussed in Chapter 6, the cry for 'liberty, equality, fraternity' was sounded forth in the French Revolution as an incipient expression of what is still to be fully realized as a conscious threefold ordering of society. The ideal of *liberty* relates to individuality, to each and every human being who is sovereign over him- or herself and develops very individual aims in life. Polar to this is the ideal of *fraternity* which is the will-to-cooperate, where the impulse for individual freedom is not the driving force. It is easy to see how this polarity of ideals can generate social conflict, the tension between what the individual wants or needs and the conformity required for society to function as a collectivity. However, between these conflicting opposites there is a mediating element which is the ideal of *equality*. Equality expresses the realization that every human being is equally free, equally individual, and is equally within the law. Equality expresses and mediates the impulses of both individuaity and collectivity, both the 'I' and the 'we'.

The threefold nature of contemporary society was recognized by the eminent German sociologist and philosopher Jürgen Habermas, who refers to three sub-systems of the total social system, and also the three life-worlds of society. These three interrelated spheres Habermas calls the economic system, the political-administrative system and the socio-cultural system, each with a separate 'control centre' in organizational terms.[239] Before Habermas the notion of society as a threefold living organization had already been developed to a high degree in the social thought of Rudolf Steiner, including how it gradually emerged in Western civilization.[240] The three spheres are differentiated but intimately and organically related such that each sphere can function and be understood only in relation to the other two. Steiner calls them the *economic, political-rights* and *cultural-spiritual* spheres.

The *economic sphere*, with respect to its function of transforming substances, is founded on the task of providing for human beings' material needs and, as defined in classical economic theory, is bounded by the availability of natural material resources. To provide for material needs, raw materials are transformed, goods are produced and distributed; this involves an immense cooperative or fraternal activity on the part of human individuals which we term 'division of labour'. It is true that in today's economic life people work within a wage productivity nexus; but this is nothing like the instinctive work relationships of a hive of bees. People, to a greater or lesser extent, *choose* to carry out a certain kind of cooperative work and do so out of a sense of interest in and care for the needs of others. It is a highly jaundiced view of human nature which insists that human beings only work together in the economic sphere for their own personal survival or benefit. Cooperation or solidarity is the defining ideal of the healthy economic sphere.

The *cultural-spiritual sphere*, with respect to its function of cultivating and realizing individual gifts and talents, of enabling creative inspiration and cognition, is polar to the economic sphere. The cultural-spiritual sphere comprises the arts, religion, education, scientific research, all of which, in one way or another, aspire toward the ideal of freedom. However, just as we cannot divide body from mind in the human constitution, so we cannot divide spiritual creative or cognitive work from its material manifestations and from the fraternal life. We do not become free in our concepts about freedom; we become free in conscious action, in deed. The work of the artists is to embody their creative impulses in movement, pigment or the word; the scientific theory seeks application. We recognize the ideal of freedom primarily in relation to the cultural-spiritual sphere but the results or influences of individual creative freedom work throughout society as a whole. In this it can be compared to the sense-nerve system of the human body which is focussed in the brain yet penetrates the body as a whole.

In society there is a third sphere which represents the mediating element between the economic and cultural-spiritual spheres; this is the *political-rights sphere*. Here neither cooperative selflessness nor free creative expression is primary; rather, what is significant in this sphere is the means by which these two can coexist in the social organization. The harmonious living and working together of human individuals is not merely instinctual as with a hive of bees; neither is it merely the result of free creative activity. Laws apply to all communal activities and individuals have certain equal rights in relation to others; the political-rights sphere is the copula, the joining and mediating element. Every right is individual (for example, the right to life) and therefore relates to the sphere in which the ideal of freedom is primary. But any right doesn't just relate to 'me'—if I have a right then every single other human individual must equally possess that right and in that way all rights are related to the sphere in which fraternity or cooperation is primary.

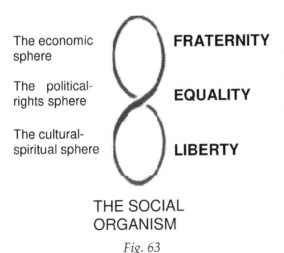

The economic sphere

The political-rights sphere

The cultural-spiritual sphere

FRATERNITY

EQUALITY

LIBERTY

THE SOCIAL
ORGANISM

Fig. 63

Our focus here is on the economic sphere but, as will be seen in what follows, economic phenomena can only be understood in a living, imaginative way in relation to the other two spheres. This is the mission of the Goethean scientific approach as applied within a tertiary education in economics: to consciously participate in economic phenomena to the point of cognitive feeling and cognitive will. Steiner termed a 'new Goetheanism' the development of the methods of Goethean life science to the study of society, in all its forms and manifestations.[241] The Goethean approach brings us to the threshold of the sphere of living form from which vantage point we see the elements of classical economics in a new light. This 'new Goetheanism' builds on classical economics but its approach and methodology is something entirely different. Classical economics and the typical university economics course today considers the economic realm as something in and of itself; in the 'living economics' of a new Goetheanism every economic phenomenon is grasped in terms of its belonging-together with the cultural-spiritual and legal-rights dimensions of society.

In the classical economics instigated by Adam Smith the economy is considered as having three parts which he called the 'factors of production'—land, labour, capital. By land, Smith meant not just the mineral earth but the waters, the oceans, flora and fauna; specifically he meant private ownership of any or all of these elements which can be used and rented for another's use. Smith saw land as just an aspect of human sovereignty and thus a dimension of the body politic—because he saw no great separation between the economic and the political. What he called 'land' is 'nature' in today's thinking—for, most fundamentally in economic life, it is nature upon which humanity works to draw the materials it needs for production. Today the enormous advances in environmental consciousness have taught us the importance of not just seeing nature in terms of private property or human sovereignty. For a start, many aspects of 'land' do not come under the category of private property—for example, gas in coal seams and fish. All natural elements are linked in a broad ecological picture—if a person cuts down timber on a privately owned property, that property is still part of a larger ecosystem which will be affected by changes to the property. Further (as will be explored in what follows) unworked nature has no economic value but most certainly has value—it has 'natural value'. Unworked nature—in any shape or form—can never be a form of capital.[242]

Nature has been liberated from the notion of 'land' as conceived in classical political economics thinking and this has come about, since the time of Adam Smith, through the continuing differentiation of society into three autonomous spheres. More specifically, the economic realm has become far more autonomous, far more emancipated from the political, than in Smith's time. Through this process it has come about that the question of nature has come more into the sphere of rights. Does human society have the right to take whatever it wants from nature in a completely unsustainable way? How we work nature, how we draw upon the resources of this planet which are in fact common to all—these are rights issues. Going even further we come to the issue of land ownership. Rudolf Steiner makes the point that sinking capital into land by owning it does not increase its value at all—it only creates a semblance of value.[243] Real value only comes about through the process of *working* that land in some economic way. Capital becomes unhealthily congested through land ownership but disappears through actual work, which means that only the work itself, the economic process, should be capitalized. The questions of whether land can be owned, who it is to be used by, are rights issues.

The second of Smith's terms refers to the element of the economic process which acts on nature productively—that is, human labour. However, his use of this term relates to the classical conception of human individuality and productivity in which labour, like land, is bound to the body politic and the wealth of the nation for which no great separation exists between the political and the economic. For the origin of classical economics we do not go as far back as serfdom when a person's labour-power was appropriated in its entirety by the ruling lord; nevertheless, Smith's view of labour is that a person's labour-power is rented for someone else's use (in other words, wage labour), just as land is rented in the traditional economic sense. Smith talked about 'productive' and 'unproductive labour', the latter—such as the work of teachers and artists—not contributing directly to the increase in material wealth of the body politic. A great deal has developed in recent times in terms of how people are employed, with the introduction of flexible working hours and so on. However, just as with the emancipation of nature from the traditional conception of 'land', so too human work has the potential for emancipation or 'liberation'— and this is indeed what we see in the threefold social organization as pictured by Rudolf Steiner.

In classical political economy, the human being is still largely a function of the body economic and body politic, a cog in the 'state machine'. Economics as a science, since its inception, has been preoccupied with working out ways to measure labour productivity based on a machine conception of work and its measurement of productivity in terms of the ratio of labour input and output is not different from how the work of a machine is measured.[244] In the threefold social organism it is in the emancipated cultural-spiritual sphere that the true value and meaning of human individuality is realized. If we take what is primary to be the unfolding of an individual's unique gifts and capacities, then this becomes the fundamental human element and labour is always in the service of this element. Writes economist Christopher Houghton Budd out of his picture of a threefold social organization:

> When we treat labour as an economic category we obscure from our vision the real relationship between capital and the individual. Instead of buying labour it would be better to capitalise the individual.[245]

In classical economics, everything is about increasing material wealth through the production, distribution and consumption of commodities derived from nature through labour. A step in the direction of perceiving the true value of the human being has been made since the 1960s in what is now known as 'human capital', which is the stock of competencies, knowledge, habits, social and personality attributes, including creativity, cognitive abilities, embodied in a person's ability to perform labour.[246] However, seeing a human being this way is still employing the same logico-mechanistic mode of thinking about labour upon which traditional economics was based. In terms of the threefold social picture the free human individuality belongs to a sphere separate from the economic: the cultural-spiritual sphere. In no sense at all is an individual 'capital'—a person is a free being and what they do by way of work in the world is a *right*—the right to freely fulfil their potential, to best utilize their capacities and gifts. Labour, in relation to the realized threefold social organism, must be seen as belonging to the rights sphere.

A living conception of economics brings forth a new picture of capital, capital being the third of Smith's 'factors of production'. Perhaps more than any of the three 'factors' it is easy to assume that capital is an economic phenomenon—but it is not. Or rather, in a realized threefold society in which each sphere has decidedly, consciously come into its own, capital has metamorphosed into a rights phenomenon. Capital in its essential forms—financial capital and capital goods—arises in the economic sphere, primarily through transformation of the phenomena of nature. As will be seen in what follows in this chapter, capital can be understood to be the outcome of a kind of distillation process; the transformation and 'distillation' of the material entities of the natural world (minerals, plants, animals) so that they come into a state of pure potential. Capital is potential; its value lies in nothing other than what it can enable to take place. Capitalism, as an economic system, takes place through the garnering of capital through ownership of the means of production and through the resultant profit. Capitalism promotes the idea that capital is an economic phenomenon which can be possessed, like a material object. Yet the real value of capital lies, not in itself, but only in what individuals can achieve *by means* of it. Capital arises in the economic sphere but comes, in its most heightened form, to belong to the rights sphere where it pertains to the cultural-spiritual sphere. Human individuals have a right to labour, to freely develop their capacities to the greatest possible extent, and they also have a right to capital to the extent that capital will allow these capacities, these creative endeavours, to be realized.

In summary, the suggestion being made in the above outline is that all three factors which Adam Smith held to be fundamental to economic science—land, labour and capital—are in truth, and in a healthy social condition, *rights* phenomena. This is not merely a matter of theory; it is an understanding which comes about through a living form of economic thinking. In relation to an orientation course of study which is preparatory to the teaching of economics proper, the method is phenomenological. It is always a matter of *seeing*—and we know from our previous study of Goethean science that this means seeing with 'the eye of the imagination'. Just one of the three factors—capital—will be explored more deeply in what follows. Through Goethean science and the awakening of cognitive feeling toward cognitive will, students have the opportunity to see and actually cognitively *experience* how capital, the fruit of the economic process, enters the sphere of rights and bears upon the sphere of cultural and spiritual life.

Cognitive Feeling (Goethean Science)

In the following discussion the meaning of polarity, metamorphosis, intensification and archetypal phenomenon are related to two aspects of the economic process—the creation of capital and the phenomenon of money. However, the kind of thinking which is developed here—a dynamic, imaginative thinking—can and should be applied to the entire organization of the economic realm, in all its forms and functions.

We place before ourselves the social organism as an image of the physical-soul-spiritual constitution of the human being. We are conceiving society as a living being, an organism in the uniquely human sense. Society is an image of what a human being *is* in its full reality. A human being is more that just organized matter; a human being is also mind and ego ('I') which we will call here 'spirit'. Thus we can picture society as existing between the poles of matter and spirit, and all social processes and forms as pertaining to this polarity. Every economic process and entity can be mentally pictured as coming into being and dying away within the living social wholeness which is polarized between matter and spirit.

Matter means tangible substance which in our bodily constitution we share with all forms of nature—mineral, plant and animal. In terms of the social organism and focussing as we are on the economic sphere, we will call the material pole 'nature' ('land' in terms of classical economics) because what is required economically is, for the most part, different forms of physical substance. Nature includes mineral ore deposits, gas and oil reserves, timber and livestock—anything of a more or less solid, physical nature. The whole of the natural realm provides the potential and stimulus for economic development.

Spirit, which we are also calling mind, is polar to nature because it is of an entirely non-physical character. The concern here is not how mind arises in the human being, or its connection to brain function and so on; the important point is simply that ideas, intentions, mental images, feelings, imaginations are non-material. Spirit is polar to matter but these two are dynamically unified, being poles of the organic whole we call the human being. Spirit is unceasingly engaged with matter because it is through the impulses of spirit (intentions, creative impulses etc.) that we *work* on nature, transforming it. Matter and spirit are not mere opposites; they represent a true polarity because they have a mediating or binding principle, a copula. Something new *arises* through the interaction of the polar principles and this is the whole realm of human social production.

a) Capital goods

The fraternal aspect of human nature is awakened and drawn out in the way we engage with the Earth economically and allow it to be transformed and enter into social life. Most essentially, we strive for sharing and coop-eration so that all aspects of economic activity take place as efficiently and productively as possible. Any finely made, beautiful product—even some-thing simple like a paintbrush—is the outcome of the cooperative actions of a large number of people, going right back down the supply chain to the producers of the raw materials.

When we speak of raw materials or commodities which have a use value, we mean that nature has come within the economic purview and a first step in transformation has taken place. Three initial steps in the economic process are as follows—from nature as such, to nature as raw materials (say, felled timber and extracted iron ore), to a good such as a hammer. Spirit (mind) works on nature as conscious inten-tion, transforming it into commodities. When we look at milled timber or iron ore, nature still 'shines out' to a large extent. When we look at a hammer we perceive that nature has receded and the 'idea' shines out; this is the hammer's value or meaning. As a good such as a foodstuff is consumed, its value decreases and its substance returns to nature; but unlike a foodstuff we call a hammer a *capital good* because it holds the potential for another form of activity. The hammer can be used to build up other social formations which have an even greater capital value—for example, a house.

Let us consider the trunks of trees in a forest. In their natural condition, unworked, they have no economic value—whether they grow on private property, in a plantation or in the wilderness. Nature is altogether outside the economic process. Trees are forms of natural creation—and we see this to be true through any form of Goethean 'delicate empirical' study which approaches the forms of nature in terms of what they are in themselves, not according to theory or intentions we may wish to place upon them. However, if the human spirit approaches them with a certain economic intention and as soon as they are worked on and usable pieces of timber are fashioned through labour, they become seed-forms of the future—that is, they become capital.

Fig. 64

Cognitive feeling enters into this process in a participatory way, through forming mental pictures of a dynamic kind, which enable us to cognitively *experience* this dynamic relationship of spirit and matter. We experience the definite form of the tree trunks—their slender height and breath, their rigidity, the colours and texture of the bark and so on; we carry out a Goethean observation on these natural forms to a greater or lesser degree. Then, moving to the planks, we observe firstly the breakdown of the natural form, similar to the way we cut up foodstuffs in order to reorganize them and digest them. This renders nature relatively formless in order to prepare the tree substance for future use by human beings. The wood loses its natural form (it becomes planks) to prepare for a new forming process out of the human spirit; this potentiality is *commodity capital*. Cognitive imagination enters into this metamorphosis and experiences *exactly* the difference between a tree trunk and a piece of timber. This rendering formless is the first step in the *intensification* of nature into the economic sphere; the timber planks are called commodities which have value as capital inasmuch as they will allow a human creative capacity to unfold in the future—namely, the capacity to shape and build.

Let us continue this intensification of nature: we observe that nature becomes progressively spiritualized within the economic sphere. Intellectually, in terms of economic theory, this may appear as facile but on the level of living thinking it is a demanding mental task to work through this metamorphosis. The 'spiritual form' of the timber planks is fashioned into a new form—a wooden plane (includes a metal blade). This more intensified form we call a capital good if it is intended to be used for further economic production. Between the raw timber planks and the fashioned object, capital is intensified but in a very specific way showing a specific metamorphosis: from natural form, to formlessness, to a form (or expression) of the human spirit, a 'concrete idea'. Now we can place these three in a metamorphic series representing an economic activity; we enter this activity and re-enact it with exact imagination. We need not have the images in front of us: we may effect the metamorphosis as an exact inner picturing to conceive the inner structure of the transformation.

SPIRIT POLE

CAPITAL GOOD: Spirit or mind realizes itself in substance (ie. a tool which allows a human gift or capacity to creatively unfold=capital). Spirit shines forth in the form of the plane and nature has largely receded (we do not see wood as wood; we see the embodied idea which is the form and function of the plane).

RAW MATERIAL: Minimally worked timber in which the substance of nature becomes a vehicle for the spirit = commodity capital.   Natural substance has become relatively formless and thus potentized. Nature still shines to large extent through the form.

NATURE: Has no economic value or potential *in itself*. The tree is the form endowed by nature; nothing else but nature shines forth in the form. Trees are just trees, not potential capital goods.

NATURE POLE

Fig. 65

We can consider the creation of a work of art such as an oil painting, to gain further experience of the spiritualization of earthly substance. The artist begins with raw materials (pigments) and with these creates the work. The process starts with the separation of different colours on the palette more or less according to the natural divisions of the circle of colours; there is then a chaotization and mixing (a rendering formless) of this order so that spirit can take hold of the substance to bring itself to expression. Through a metamorphosis and intensification the substance-nature of the pigment disappears in the becoming of the work. In the completed work pigment-nature has receded and *we only see the idea*. When we look at a painting by Chagall depicting a floating cow with a parasol, we do not see pigments but only the idea, the meaning. Works of art, like capital goods, are forms of concrete spirit. As with the formation of the capital good, we re-enact the process of creation of the painting with our exact imagination.

b) Money and Capital

Let us return to the economic intensification of nature into capital but now thinking of it specifically in terms of the metamorphosis of *values*— that is to say, the metamorphosis from natural value into economic value.

This time we can consider the example of cowrie shells which are, like the trees, creations of nature. It is the same as with trees: whatever human intentions may be, cowrie shells in their natural condition are valuable in and for themselves, without being justified in terms of human values. A Goethean 'whole animal' study can be connected with this economic research, a study through which we learn to recognize the inner completeness of the cowrie's form. Such a study would serve the purpose of guiding our imaginative thinking into this first step of comprehending the formation of money. Thereby it becomes possible to see that, in its natural state, the cowrie stands entirely outside the economic order.

We have observed through exact imagination that, in shaping wood into planks, we form a commodity and confer on it a condition of potency or seed-value (commodity capital) which relates to its potential for further formation into a capital good. In exact imagination we moved from one to the other, inwardly experiencing the transformation. In the case of the cowrie we do the same—not rushing the process but with cognitive feeling moving forwards and backward in the sequence to discover the lawfulness of the transformation. The metamorphosis of values is expressed in three stages—from the natural condition of the cowrie (the natural value), to the commodity value of the raw material (collected shells), to the economic value of a saleable good (the finished jewellery item). Labour, economically speaking, is what works on the natural forms of the cowrie, in the first place just by collecting them. The collected pile is the 'formlessness' which is the breakdown of the natural order, rendering the shells seed-forms for future creative action.

To comprehend the commodity value of the shells is an exercise in cognitive imagination. We picture the degree of labour expended to find, collect and prepare the shells. Into this picture we bring our knowledge of the form and character of the shells—their size, durability, colour and shape as well as their potential usefulness for specific creative works. Spirit may further imprint itself on the commodity by the labour of crafting the shells into a piece of jewellery. As with a work of art, nature recedes and 'the idea' shines forth. Thus, the further task is to come to a cognitive imagination of the economic value of the finished good. We build an inner picture of the appearance of the completed jewellery, its craftsmanship, its beauty, its functionality, the time spent in producing it and so on. All these elements are gathered in a vivid and comprehensive cognitive imagination of the piece of jewellery and we are now in a position to come to a meaningful assessment of its economic value or price, if this is necessary within the economic system of a particular society.

Thus we can say: through work and creativity nature moves toward the realm of the spirit and achieves an ever-higher economic value, right to the end of the production process. We re-enact this intensification process in our imagination, just as we do a plant metamorphosis in Goethean science studies. None of these values is sense-perceptible and thus quantifiable; we can perceive them only with the eye of the imagination.

SPIRIT POLE

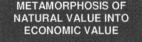

METAMORPHOSIS OF NATURAL VALUE INTO ECONOMIC VALUE

NATURE POLE

Fig. 66

In certain prehistoric cultures including those of Africa, India and China, the collected cowrie shells assumed another kind of economic value—*a monetary* value. In these cultures the cowrie shells serve as a 'commodity money'—that is, instead of direct exchange in trading (barter, where the barterers exchange some shells for something the pair mutually consider to be of equal value—say, a ceramic pot) the shells acquired a value as a unit of indirect exchange; that is, as a currency.[247] These shells are suitable as units of currency because they are small and easily handled, very similar to each other, durable and attractive. The first step we have already worked with: the activity of spirit whereby the cowrie in its natural state (natural value) is transformed into the collected and gathered shells (commodity value). But now the shells take a different direction; now spirit imprints on the collected shells a value of another kind which remains even when the shells are passed between members of a community.

To follow this second direction we need to look at the same shell with a different eye. Let us place one cowrie shell before us and observe, as a purely physical phenomenon, that it is impossible to know whether it is a commodity or a unit of money.

But if we now observe a bone-carved imitation of a cowrie shell, used in China around 3,000 years ago, we perceive how spirit has taken hold of the shells (see Fig. 68). This bone carving is a *symbol* which stands for the monetary value of the shells. The monetary value has become emancipated from the shells-as-commodity. Nature has receded and the spiritual (or we could say mental) value now shines forth as a social meaning, which is the monetary value.

Fig. 67

We can now place next to each other one shell-as-commodity and one shell-as-money, and study each phenomenologically—looking through the visibly appearing form into the meaning of which the visible form is the expression. This meaning is the value.

Fig. 68

Now each kind of value—natural, commodity and monetary—can be experienced by moving from one to the other, developing a mental picturing of one and transforming it into the next, round in a circle in either direction for the sake of the exercise in fluid metamorphic thinking. It needs time to be dwelt on deeply—for, while logically this is something quite straightforward, as a task of cognitive imagination it is considerably more difficult and unfamiliar (see Fig. 70).

As we have seen with the shell money, the first monetary forms are either natural forms or imitations of them. Money arose out of nature. We see the vestige of this connection with nature in metal coins from the Near East which depict animals and also nature deities—in the examples here the Greek god Athena and the owl (see Fig. 71).

SHELL-AS-COMMODITY: The value is a function of the form of the shell and how it can be used — for example in jewellery making, body decoration or functional items such as a pot trivet. The value is its specific (and limited) potential for such creations.

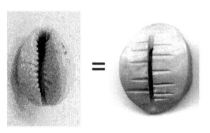

SHELL-AS-MONEY: The value is emancipated from the form and function of the shell; it is 'free'. The value therefore is a potential which is not limited. The shell becomes the medium for exchanging any items which are of equal value. The replica carved in bone has the same monetary value.

Fig. 69

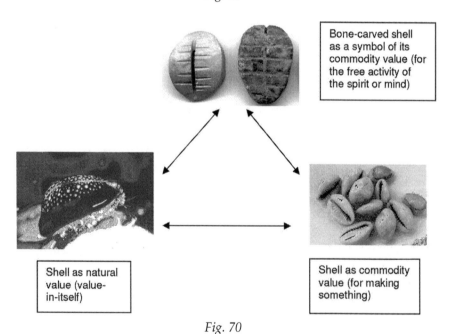

Bone-carved shell as a symbol of its commodity value (for the free activity of the spirit or mind)

Shell as natural value (value-in-itself)

Shell as commodity value (for making something)

Fig. 70

However, with the rise of aristocratic forms of governance in ancient Rome and later, coins lose this link with nature and typically depict the rulers and images from the Empire. Nature only still shines through in that the coins are made of natural substances—generally silver or gold—and to that extent still have vestigual commodity value. When money becomes entirely detached from the value it has as natural

Fig. 71: Tetradrachm (= 4 drachma) from Athens about 450 BC. Athena on the obverse, owl on the reverse side.
Coin of Emperor Constantine the Great 307-337 AD.

substance then we have *fiat money* issued by the ruling authority as legal tender. We have seen how in ancient China an imitation cowrie shell stood symbolically for the monetary value of a cowrie shell; in that civilization around the tenth century AD there occurred the first known use of paper money and this became general practice in other countries in later centuries.

In the history of money we thus see the gradual emancipation of money from nature through the work of the spirit. Spirit firstly takes hold of natural forms and substances and endows them with value (commodity money). Monetary value is of the nature of spirit; it 'speaks' directly to the needs and intentions of the human spirit. Value is *potentiality*. Gradually money becomes emancipated from natural form and substance entirely through this intensification towards spirit—it becomes numerical value only, or digital money, stored and transmitted through computer technology. In this process it also becomes emancipated from particular regions or national boundaries and becomes instantly accessible to the global economy.

We can now follow this metamorphosis and intensification with exact imagination, experiencing as a dynamic inward picturing how nature progressively transforms into spirit in the evolution of money.

SPIRIT POLE

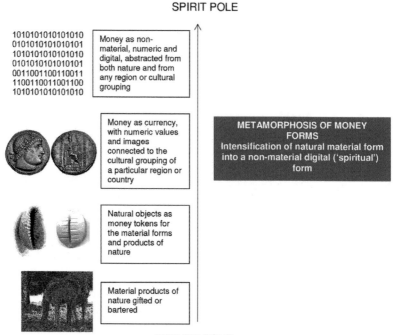

1010101010101010 0101010101010101 1010101010101010 0101010101010101 0011001100110011 1100110011001100 1010101010101010	Money as non-material, numeric and digital, abstracted from both nature and from any region or cultural grouping

Money as currency, with numeric values and images connected to the cultural grouping of a particular region or country

METAMORPHOSIS OF MONEY FORMS
Intensification of natural material form into a non-material digital ('spiritual') form

Natural objects as money tokens for the material forms and products of nature

Material products of nature gifted or bartered

NATURE POLE

Fig. 72

Let us consider the digit 4, standing for four units of any currency. A numeral is something entirely emancipated or abstracted from nature—it has only a mental (spiritual) character. We have already explored phenomenologically the cowrie shell currency and discovered that the monetary value must be expressed in these kinds of terms: the free potentiality of the human spirit to fulfil its needs and intentions in the future. Such is the nature of money as capital. What then does the numeral 4 exactly stand for in relation to this free potentiality? Money becomes emancipated from nature and freed from the economic process which is 'nature turned into commodity turned into goods'. Money acquires meaning in relation to the *rights sphere* of society. Money is a social phenomenon which means that its value and power must be seen in terms of community relationship. *Numisma* is the Greek word for money, meaning custom, consensus, convention; the value of a unit of currency comes about as a social agreement. As Christopher Houghton Budd puts it:

> When people arrive at a consensus they are not involved in economic processes, but in 'rights' processes. Money belongs to the rights life, it enables the rights life to permeate economics.[248]

It is through a social agreement that the potentiality of the human spirit to fulfil its needs and intentions becomes defined in an exact way through the numeral 4. The monetary value of 4 gives an individual the *right to purchase* anything which is socially agreed to be equivalent to the monetary value of 4 units.

From these considerations we see that money stands out as an *archetypal phenomenon* insofar as its appearance within the life of contemporary society sheds light on the archetypal threefold nature of the social organism. Money is primarily a rights phenomenon; it only has meaning in terms of the *political-rights sphere* from which every person gains equal right to use it as legal tender (ideal of equality). From there it enters into the *economic sphere* as the medium of exchange; it is the driving force which allows for the cooperative cycling of production, purchase and consumption (ideal of fraternity). And money as legal tender also permeates the *cultural-spiritual sphere* of society because it 'speaks' to the human spirit as free potentiality, the potentiality which, in many different ways, can allow human creative potentials or needs to be fulfilled. In this last sense money has meaning in relation to individual freedom (ideal of liberty). Rudolf Steiner calls money a *Dreigliederung*, a threefold membering or threefold articulation.[249]

We now can work our way with fluid imaginative thinking into the reality of money as archetypal phenomenon. The important thing is to move beyond the merely logical to a dynamic, living conception of money, in just the same way we would proceed if we were entering with cognitive imagination through the outward physical appearance or form of a work of art into its meaning or significance.

c) Production, consumption, exchange, capital

We have recognized that capital, price and value have no sense-perceptible existence in economics, that these are realities which can only be perceived with the eye of the imagination, as dynamic inner pictures. In earlier chapters the expressions 'cognitive imagination' and 'cognitive feeling' have been used, indicating that more of the human being than just the logical intellect must be involved in order to think the reality of living form. The development of these new 'organs' of economic understanding is what we are primarily concerned with here.

Through Goethean colour studies we gain experience in cognitive feeling and this developed faculty is then applied to think the living, dynamic reality of economic production, consumption, exchange and capital. Yellow, blue, green and magenta are employed here to facilitate entry into the cognitive imagination of economic phenomena; the suggestion is not being made that colours represent exact correspondences with these phenomena.

Through the Goethean study of colour we learn that through a sense-perceptible form (a colour) we can 'see' a dynamic quality which is not sense-perceptible yet nevertheless real (see Chapter 5). Blue retreats into itself; it carries the gesture of depth and inwardness. We 'find ourselves' in blue, the ego comes to sense itself but in a way which draws it to an infinite depth (see Fig. 73).

Yellow radiates in all directions; it carries the gesture of infinite outwardness. We 'lose ourselves' in the diffuse radiance of yellow—in a sense we die to self (see Fig. 74).

Thus, in terms of gesture, yellow and blue express a polarity and opposition. As we have seen in Chapter 5, all the other colours in the colour circle represent the mediation of this opposition, relating the two either through harmonization (green) or intensification (towards magenta).

Economic phenomena like commodities and goods, too, are the outward face of living dynamic qualities or gestures which are not sense-perceptible; these are the forces at work in the living social fabric. Now that we have explored the economic process through exact imagination as far as the production of a good (the piece of jewellery), we can take a broader look at the process of production, penetrating with cognitive feeling to bring light upon its characteristic gestures. In the intensification of nature into the realm of the spirit we are dealing with the first half of the economic process—with spirit involving itself with nature and creating economic value through productive action. The economic product formed through this intensification is created in relation to the specific needs of others and this gives the process of production a peculiarly *outward* 'yellow' gesture. The fact that, on sale, a financial return to the producer may occur, takes nothing away from the essential selflessness of the act of production. In the economic sense of the transformation of nature, we do not produce because *we* need to produce for ourselves but because *others* have needs.[250] Pictured as an inner imagination, every product gestures out in all directions towards specific needs of human individuals in society.

When a good such as an item of jewellery is produced, it arrives at a threshold which is its point of sale. As long as it stays at this point, it exists in a condition of unresolved tension. The product exists at this economically intensified point only so that it may be sold and this is what creates the whole movement of the economic process. In a healthy economy the outward, proffering gesture is resolved at the moment of sale when the activity of consumption begins to take place. Consumption carries the opposite gesture to production, a 'blue' gesture; it is an inward, self-orientated gesture whereby the product is eventually used up. It is not possible to consume for anybody else; consumption is essentially self-orientated. In

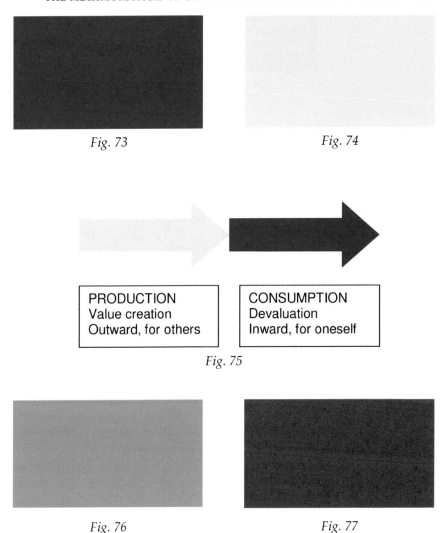

Fig. 73   Fig. 74

PRODUCTION	CONSUMPTION
Value creation	Devaluation
Outward, for others	Inward, for oneself

Fig. 75

Fig. 76   Fig. 77

consumption nature which has been intensified towards the human spirit through labour now begins to die away and the values which have been built up to the highest point when the production is placed 'on the market' are more or less rapidly negated (see Fig. 75).

Consumption is thus a process of devaluation and return to nature; eventually everything produced will 'turn to dust'. With foodstuffs this is very rapid and much longer for durable products. The product itself devalues materially but, depending on the product and the way it is used, it may lead to an increase of value on another level. We have already seen how a capital good like a wood-working plane may be used constructively whereby value is enhanced within the process of production—for

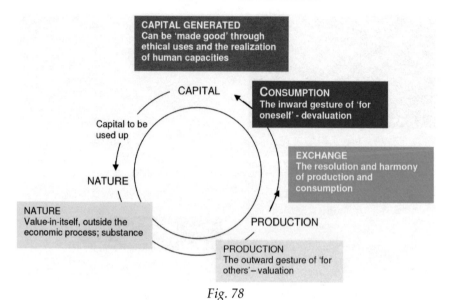

Fig. 78

example, by making toys. Foodstuffs too, even though rapidly consumed materially, provide human beings with the sustenance they need for all manner of physical and mental activities. These activities, if applied to the economic process, lead to further cycles of production and consumption.

The first step is to inwardly *experience* the polar gestures of production and consumption, valuation and devaluation, and the way they are held in an equilibrium in a healthy economy due to the fact that products are made to meet specific needs—in other words, to create a balance of supply and demand. The gestures of yellow and blue approximate the polarity of the two economic gestures—outward and inward. Together they form a cycle of activity, a reciprocating activity, an arising and passing away, thus bringing to light the organic nature of the economic sphere.

The event of a product passing from the hands of the producer into those of the consumer is called *exchange*. Exchange resolves the tension of the product on the market; its gesture is that of the harmonization—it is neither outward nor inward in gesture but is the union and balance of production and consumption. We can employ the gesture of the colour green as an approximation of this event, green being the harmonization of the poles of yellow and blue.

No modern economy is based on a system whereby products of an equivalent value are merely exchanged or bartered. The transition between production and consumption is normally mediated by money. As we have seen above, the value of money is not attached to any particular good or service—it is emancipated from the economic realm and has transformed into a rights phenomenon whereby it can permeate the economic sphere as the medium of exchange. Money can be exchanged for any different good or service at will; beyond its monetary value it has non-specific potential. The profit from the sale of a product becomes available as capital to the producer for use in different ways; as purchase capital within the economic sphere or loan or gift capital within the cultural-spiritual sphere where it can support all manner of creative endeavours. The creation of capital means that, even though the material or commodity value of the good falls away, the capital value of the product may be preserved and enhanced through the appropriate use of the capital.

We can employ the colour magenta to approximate the inner gesture of capital. Magenta, like green, expresses the union of the polarity of yellow and blue—it is neither one-sidedly outward nor inward in gesture. However (as discussed in Chapter 5) green harmonizes this polarity whereas magenta intensifies it, having both the outward strength of red and the inward concentration and depth of purple. It has a regal, noble quality and was traditionally used on the attire of sovereigns (see Figs. 76 and 77).

Production and consumption express the form of a cycle between spirit and nature because a product raised up out of nature inevitably, naturally, falls back into nature. No such inevitability is connected with the use of capital. The wholeness and health of the economic sphere on this level is an achievement of consciousness and a moral or immoral deed. Funds can be frittered or gambled away or invested and manipulated in ways which increase their quantity without producing anything of value within the social realm. On the other hand, the capital can be 'made good' through usage which promotes healthy growth processes in society and which supports the nurturing and enlivening of the Earth through everything that comes under the name of sustainability—for example, through ethical investment, renewable energy, and organic-ecological architecture. Capital, in innumerable ways (and when it isn't locked up in land), can be made good through allowing human individuals to realize their capacities. The usage of capital for purposes which nurture, preserve and ennoble carry something of the regal gesture of magenta with its outward strength and inward concentration.

In summary, we perceive four fundamental gestures of economic life—production, exchange, consumption and capital. After exchange the modern economic life gives expression to two opposing tendencies: firstly, through production, exchange and consumption, there is finally a devaluation and the destruction of physical form and the return of substances to nature. Opposing this, yet united to it, is the production of capital which allows human capacities to unfold, creative impulses to be realized. Here constructive, creative, fructifying forces are at work.

- PRODUCTION: From the intensification of nature into concrete spirit (commodities and products) we experience production with its essentially outward gesture of 'for others', meeting the needs of others.

- CONSUMPTION: From the highest point of the productive process, the product begins to 'die' back to nature in the process of consumption with its essentially inward gesture of 'for oneself'. *Production (valuation) followed by consumption (devaluation) forms an organic cyclic process uniting nature and spirit.*
- EXCHANGE: Neither outward nor inward in gesture—the union and resolution of production and consumption.
- CREATION OF CAPITAL: Through exchange capital comes into being which may be used creatively and responsibly to develop and enliven nature—production (valuation) can be given additional value by 'making good' the capital. *The organic form of the economic order created consciously out of free action.*

The four economic gestures in relationship now may be represented pictorially using the four colour gestures as approximations. Viewed together imaginatively they form a picture of the economic sphere in its wholeness or organic unity and, as such, constitute the foundation for a more developed education in economics at the tertiary level (see Fig. 78).

d) The production and use of free capital

We have already looked phenomenologically at the intensification of nature into spirit and the production of capital goods such as a woodworking plane. In the capital good, spirit realizes itself in substance as an embodied idea (the plane) which allows a human gift or capacity and a particular creative impulse to unfold in a very specific way. In the economic sphere such a capital good comes into being through the specialized, cooperative work of many people—including tree-fellers, timber-mill workers, tool-designers, factory-machine designers and machine operators, iron-ore miners and smelters, blade manufacturers.

Closer to the 'nature pole' we may picture to ourselves the single human being who, in intimate physical and spiritual connection with nature, fashions a simple tool such as a stone axe from the raw materials of nature—from wood, stone and grass fibre—which he has obtained. This person makes this tool by himself for himself—thus it is not part of any economy. However, if he barters the axe (exchanges it for some equivalent item) then at that moment an economy comes into being. In this primitive, largely undifferentiated society there is no money— and we have seen above that the coming-about of money is connected with the differentiation of society into spheres of economy, legal-rights and cultural-spiritual life. Money permeates all three spheres but, as legal tender, it is primarily a rights phenomenon. In modern, highly differentiated forms of society an immense division of labour is involved in the economic life, the result of which is highly efficient production processes and large-scale profit-making. Now money is the medium of exchange and, through profit, free capital comes into being.

Consider the growth of a plant; in its seed stage the plant belongs primarily to the Earth and its dark enclosing spaces; in the vegetative stage with its sprouting, spreading and spiralling rhythms of leaf forms it belongs primarily to the realm of water, air and light. With the flower the plant is given over to the greatest degree to light and warmth of the Sun and this is revealed in the flower's colours and fragrance, the sweetness of the nectar, the cup-like receptive space of the blossom into which an insect or golden pollen descends from light-filled heights. The flower belongs primarily to the sphere of the Sun. This is best worked with as an imagination—that is, as a Goethean experiential study moving from cognitive feeling towards cognitive will. It could involve painting exercises, or poetic creativity, in order for students to grasp how, in the metamorphosis of a living being, new forms of that being are assumed in different realms.

Similarly, capital expresses stages of transformation and distillation. The plant is not an analogy for the creation of capital but the living thinking required to properly understand plant metamorphosis allows us to enter into metamorphosis in an entirely different sphere of living reality, the sphere of economics. Such is the task of a 'new Goetheanism'. At the stage of *commodity* capital still belongs primarily to the Earth and to the economic sphere of the social organism. As *capital good* (such as a hammer, computer or any machine which comes under what is meant by 'means of production') it is still part of economic production yet more orientated toward the cultural-spiritual; for with the capital good, nature has receded and the 'idea' shines forth most clearly (see above). In its most distilled form as *pure capital* it expresses indeterminate potential, and in this state belongs primarily to the rights sphere, orientated towards the cultural-spiritual sphere. This is because capital as spiritual potential 'speaks' principally to the human creative spirit, the individual capacities and creative intentions which can be realized through capital's potential. This entire transformative process needs to be brought into a mobile, living inner picture.

Capital, thus, is a creation *of* the human spirit *for* the human spirit. What does 'human spirit' mean here? It is every idea, intention, aspiration and creative impulse through which individuals seek to realize their capacities and gifts. In very many ways, on many levels, capital is required for individuals to bring about in the world what is most unique to themselves. Every individual requires capital and has a right to capital, in order to realize their capacities and gifts. Capital is necessary for everything, from accommodation, to nourishment, to equipment. This individual unfoldment and creative realization is the meaning and essence of the cultural-spiritual sphere of society. It includes the work of artists, scientists and educators most obviously, but just as much mothers and factory workers. Importantly, it includes business entrepreneurs, for the creative work of the entreprenear begins as an inspiration or impulse within the cultural-spiritual sphere and then works over into the economic. Agricultural, technological, manufacturing impulses all have their origin in the cultural-spiritual sphere. The cultural-spiritual life is continuously fructifying the economic life.

In the diagram below we see the magenta gesture of capital as pure potentiality within the rights-legal sphere, distilled from nature in the economic sphere through the impulse of spirit, meeting the creative aspirations which characterize and define the cultural-spiritual sphere of society.

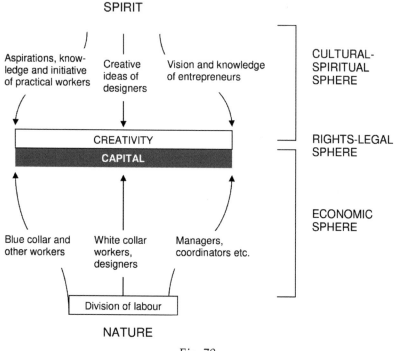

Fig. 79

Human individuals have the right to capital, for it is distilled from the common Earth. No matter how ingrained the idea is that capital is an economic phenomenon, it is not. In connection with the worldwide and highly topical debate concerning a 'universal basic income' we see, although inchoate, the dawning recognition at the level of political and legal thought that capital is indeed a human right. To say that capital is the asset of the whole community means something entirely different from saying that the whole community of individuals *owns* this capital. According to Marxist thought, the aim of socialism and eventually communism is to remove capital—meaning capital goods (the means of production), land and financial capital—from the hands of the few and place it into the hands of the many in order that they can control it for their mutual benefit. Marxism, just like capitalism, conceives capital only economically, fraternally, treating it as a commodity which can be possessed—either by few or many. But capital, as we have seen, is not an economic phenomenon at all; it arises in the economic sphere but comes in its most heightened form to belong to the rights sphere where it pertains to the cultural-spiritual sphere. No person has to do anything to win the right of access to capital; capital cannot be the compensation for labour in the economic sphere as if that labour could be purchased like one purchases a work-producing machine such as a motor or a robot ('wage' is the name we have for that purchase of labour). Capital is a right, meaning—like every right—it is accorded to a person simply by virtue of them being a human being. And it is the task of the political-rights sphere to ensure this right is accorded. [251]

Every person has the equal right of *access* to capital but that doesn't imply that every person has to receive an equal amount of capital; every person has different needs and different capacities. Capital is formed by the human spirit for the realization of the human spirit and the criteria and method of judgement about how capital is to be distributed and used must therefore be a determination of the cultural-spiritual sphere, by those who actually work within that sphere. Steiner writes:

> When [an] individual is no longer willing or able to direct his abilities to the use of capital [at the service of the community], this use must be transferred to another person of similar abilities. It will not be transferred by state prerogative or by economic power, but by finding out, on the strength of the training acquired under the free spiritual life, which persons will make the most suitable successor from the social point of view.[252]

Those who live and understand human freedom and the requirements of the free individuality must surely be those whose task is to administer this aspect of social life. Banks have a vital role to play in the distribution of capital because banks are 'rights organs', mediators between the economic and cultural-spiritual life. But banks themselves are not the only determiners of how the potential of capital is to be realized—or, at least, they do so only through conversation with those in the cultural-spiritual sphere who are directly involved with the needs and creative capacities of free human individuals.

Perceived in this way we understand capital as something far more meaningful for human society than book-keeping alone reveals. Capital represents the creative potential of the whole human community; it has been brought about over millennia, through a vast effort, by means of an immense division of labour, such that it can now be made available to the kind of free human individuality which has emerged through the evolution of consciousness. To keep capital free in social life means to allow it to 'live' out of its essential magenta gesture which speaks of the nobility and potentiality of the individual human spirit. Writes Christopher Houghton-Budd:

> Capital belongs to and gravitates to the spirit in us, not to our social standing, ancestry or power base. And the spirit is at work in everyone. To the degree that this becomes conscious, everyone will realize the need for capital. Society in the future will come to see this. Then every single individual will be capitalised: not just a few, with the rest cast into wagery. The great promise and potential of capital is that it always seeks the new in human beings, the uniqueness of their spirit.[253]

We can say that this promise of capital will depend to a large extent on an imaginative form of education which deals with the origin and meaning of capital in the social organism. Such is the task of economic studies within the universities of the future.

Cognitive Will (Artistic Practice)

The disappearance of nature in economic production:
This exercise is to develop a powerful inner experience of how the spirit works on nature in the production of commodities and goods. A rough lump of modelling clay is thoroughly explored in the hands to appreciate its 'natural' properties; it is simply part of the mineral substance of the Earth. It can at least be imagined that this material has value only in-itself, a natural value, not in relation to any economic intention. After this the clay can be fashioned into an item which might be a decorated pinch pot or an animal sculpture (items which could have economic value); as this is slowly and carefully carried out the attention should be on how the clay as substance 'disappears' as it transforms into the form of the work, so that what we see in the end is more the idea of the work than the clay itself. The idea subsumes the material. The aim is not so much to produce an excellent work of art but to experience metamorphosis of 'substance as substance' into 'substance as the vehicle for the revelation of spirit'. The same thing could be worked with as a painting, starting with careful observations of the form and colour of the separate pigments.

Economic gestures:

This exercise relates to the distillation and transformation of capital as discussed above, under the heading 'The production and use of free capital'. As indicated there, a living understanding of this process can be cultivated through artistic studies of plant development. On a large sheet of cartridge paper and using water colours, the wholeness of the plant being can firstly be brought to expression as the polarity of Sun and Earth. Sun is expressed as a thin but radiant wash of yellow paint which starts in the top corners and forms a downward-facing chalice in the top part of the paper; Earth by a denser blue which extends from the bottom corners and forms a terrestrial surface. Together the colours form an oval enclosing space in which the plant develops 'between Earth and Sun'. Into the blue below a small dark blue seed is 'planted' and from it roots, stem and germinal leaves develop, giving expression to the watery, airy realm. Near the top the leaves become thinner, pointed and contract back into the stem, as the reality of the flower 'approaches' from the realm of Sun—warmth and light. The flower, or flowers, are formed in the top parts of the oval; the plant now belongs to the Sun (see plant metamorphosis, Chapter 5).

This plant study can lead directly to an artistic economic contemplation which might best be carried out through poetic languaging. The economic process of capital formation takes place between the polarity of 'nature' and 'spirit' and within the threefold social wholeness. This sense of social wholeness is firstly brought to expression. The process begins with the forms of nature and 'natural value' (not economic value). Into the economic processes arises the commodity value of raw materials which are still close to the Earth pole; then the capital value of capital goods which have specific potentials or creative possibilities within the economic process and are close to the realm of culture and spirit. Nature has receded and 'the idea' (spirit) shines out. With pure capital we are in the realm of spirit and unlimited creative possibility; nature is distilled into spirit.

Destructive and creative forces in the economic sphere:

Exercises in metamorphosis through cognitive will, using clay, can develop the sense of destructive and formative forces in the economic life. Firstly, a lump of modelling clay is worked and explored in the hands, sensing it as a raw material and close to the natural condition. Now a sphere is carefully and slowly formed by the palms pressing in from all sides—not by rolling on a table. This sphere can be imagined as the creative fullness of a completed good or product, at the culmination of its production process and at the point of sale.

After exchange the good is consumed and the gesture of this destructive force is expressed by working into the sphere with the thumbs; substantial fullness becomes emptiness. Carefully and deliberately, the concave sphere is transformed into a concavity; the complete-unto-itself or self-sufficient quality of the sphere transforms into a cup form with a hollow, receptive quality. A negative gesture has created a positive, and this positive is a potentiality, a receptivity, which relates it to the gesture of capital. Destruction and creation are bound up with each other—this is the insight which can be developed through this exercise. If a good is used up (consumed) creatively, wisely and ethically, then a greater good can be the result.

Notes

1   See in particular Thomas Kuhn, *The Structure of Scientific Revolutions*.

2   J. Naydler (ed.), *Goethe on Science*, Floris Books, Edinburgh, 1996, p.78.

3   'Dark Satanic Mills' is from the poem 'And did those feet in ancient time' by William Blake, often interpreted as referring to the destruction of nature and human relationships in the early Industrial Revolution.

4   M. Heidegger, *The Question Concerning Technology and Other Essays*, Harper and Row, NY, 1977, p.42.

5   R. Steiner, *A Modern Art of Education*, Rudolf Steiner Press, London, 1972, p.44.

6   R. Steiner, *West and East: Contrasting Worlds*, The Rudolf Steiner Publishing Co, London, p.42.

7   From 'The Fairy Tale of the Green Snake and the Beautiful Lilly' by Johann Wolfgang von Goethe.

8   R. Steiner, *The World of the Senses and the World of the Spirit*, Rudolf Steiner Press, Forest Row, 2014, p.15.

9   W.T. Reich, 'History of the Notion of Care,' in *Encyclopaedia of Bioethics*. W.T. Reich (ed.), Simon & Schuster, NY, 1995, pp.319-331. Also: www.care.georgetown.edu/Classic%20Article.html

10   See M. Heidegger, *Being and Time*, (trans. J. Macquarrie & E. Robinson), Basil Blackwell, Oxford, 1973, pp.225-230. Implicit in care, in the way Heidegger analyses it, are different life positions. One is anxiety or worry (*Besorgnis*), which relates to the many aspects of our struggle for survival. Care in the sense of 'taking care of' is close to what we mean by concern; that is, catering to the needs of others (*Besorgen*). This he contrasts with solicitous care (*Fürsorge*) meaning tending to others with consideration and forbearance, nurturing, caring for the Earth and for our fellow human beings as opposed to merely providing for their needs.

11   R. Steiner, *Freedom of Thought and Societal Forces*, Steiner Books, Great Barrington, 2008, p.98. The fact that Steiner stands well outside the academic sphere in his consideration of physical and spiritual realities seems to be the reason that, even if one just quotes him, one can be branded a follower rather than someone who simply has made reference to his ideas. In such a unscholarly state of affairs scholarship becomes impossible. For example, the academic author Astrida Orle Tantillo makes a special point, in the preface of her book *The Will to Create*, of alerting her readers to the 'problem' of Rudolf Steiner and his 'followers' who often write on Goethean science. By 'follower' she appears to mean anyone who even just quotes or mentions Steiner in their writings. Tantillo refers to Henri Bortoft who does just that in his important work on Goethean philosophy, *Goethe's Way of Science*. She says that these 'followers' tend to treat Goethe's

science as 'a kind of mysticism or religion' and to look in Goethe's texts for 'messages of personal/spiritual guidance and fulfilment'. Tantillo wants to prove her point by going on to quote briefly from Bortoft and also from authors Mark Riegner and John Wilkes. Riegner and Wilkes, she reports, talk about coming to the experience of 'the idea within reality' through Goethe's precepts. Bortoft, she indicates, speaks of a new kind of education possible through Goethe's science which activates 'a different mode of consciousness which is holistic and intuitive', thus helping to bring about a 'radical change in our awareness of the relation between nature and ourselves'. Meanwhile, in the body of her own text (pp.37-8), Tantillo expresses no difficulty with Goethe's own views on a spiritual kind of consciousness. She writes: 'Throughout his *Theory of Colors*, Goethe discusses both the *physical eye* (the organ of sight) and the *spiritual eye* (the inner organ of intuition)'. (A.O. Tantillo, *The Will to Create: Goethe's Philosophy of Nature*, The University of Pittsburgh Press, Pittsburgh, 2002.) H. Bortoft, *The Wholeness of Nature: Goethe's Way toward a Science of Conscious Participation in Nature*, Lindisfarne Press, Hudson, 1996; essays by M. Riegner & J. Wilkes in *Goethe's Way of Science: A Phenomenology of Nature*, D. Seamon & A. Zajonc (eds.), State University of New York Press, NY, 1998.

[12] Elsewhere I have developed the pathway of Goethean science according to four elemental stages, the middle two 'water' and 'air' stages representing the dissolution which prepares for the advent of creative cognition. See N. Hoffmann, *Goethe's Science of Living Form: The Artistic Stages*, Adonis Press, Hillsdale, 2007.

[13] Quoted in A. Zajonc, 'Goethe and the Science of his Time,' in *Goethe's Way of Science*, op. cit., p.17.

[14] These thoughts of Goethe were written down at the time by the Swiss theologian G.S. Tobler; see Nicholas Boyle, *Goethe: The Poet and The Age: The Poetry of Desire*, Oxford University Presss, Oxford, 1992, p.339.

[15] From F.W.J. Schelling, *Philosophie der Mythologie: Volume 5, Schellings Werke*, C.H. Beck, 1984, p.3.Translation by N. Hoffmann.

[16] As Henri Bortoft writes: 'The intelligibility of the colours in themselves disappears in the analytical approach, and what is left seems to be merely contingent. It is no answer to be told that the order the colours appear in is the numerical order of their wavelengths, and that red has the quality of red because its wavelength is seven-tenths of a millionth of a metre, whereas violet has the quality of violet because its wavelength is four-tenths of a millionth of metre. There is simply no way in which these qualities can be derived from such quantities.' H. Bortoft, *The Wholeness of Nature*, op. cit., p.48.

[17] Two who have engaged in this debate are D. L. Sepper, *Goethe contra Newton: Polemics and the project for a new science of colour*, Cambridge University Press, Cambridge, 1998, and H. Bortoft, *The Wholeness of Nature*, op. cit.

[18] See R. Steiner, *Nature's Open Secret: Introductions to Goethe's Scientific Writings*, Anthroposophic Press, Great Barrington, 2000.

[19] Goethe writes about Kant's idea of the *intellectus archetypus* in his essay 'Judgment through Intuitive Perception', in J.W. von Goethe, *Scientific Studies*, (trans. D. Miller), Suhrkamp Publishers, NY, 1988, p.31.

[20] J.W. von Goethe, *Italian Journey*, Penguin Books, London, 1970, p.38.

[21] Ibid., p.388.

[22] Ibid., p.168.

[23] Ibid., p.258.

[24] Ibid., p.153.

[25] F. Nietzsche, *Thus Spake Zarathustra*, Section 38.

[26] J.W. von Goethe, *Scientific Studies*, op. cit., p.39.

[27] R. Steiner, *The Science of Knowing*, Mercury Press, Spring Valley, 1988, pp.87-88.

[28] Quoted in G. Sutherland, 'A Trend in English Draughtsmanship,' *Signature*, III (1936), pp.7-13.

[29] See Kant's *The Critique of Practical Judgment*.

[30] C. Holdrege, *The Flexible Giant: Seeing the Elephant Whole*, The Nature Institute, Ghent, 2004, p.4.

[31] C. Holdrege, *The Giraffe's Long Neck*, The Nature Institute, Ghent, 2005, p. 91.

[32] R. Steiner, *The Science of Knowing*, op. cit., pp.92-100.

[33] A.T. Kronman, *Education's End: Why Our Colleges and Universities Have Given Up on the Meaning of Life*, Yale University Press, New Haven, 2007; B. Readings, *The University in Ruins*, Harvard University Press, Cambridge, 1996; P. Smith, *Killing the Spirit: Higher Education in America*, Penguin Books, Harmondsworth, 1990; A. Bloom, *The Closing of the American Mind: How Higher Education Has Failed Democracy and Impoverished the Souls of Today's Students*, Simon & Schuster, NY, 2012; F. Donoghue, *The Last Professors: The Corporate University and the Fate of the Humanities*, Fordham University Press, NY, 2008; W. Chapman (ed), *The Western University on Trial*, University of California Press, Berkeley, 1983.

[34] J. Pelikan, *The Idea of the University: A Reexamination*, Yale University Press, New Haven, 1992, p.37.

[35] N. Lobkowicz, 'Man, pursuit of truth, and the university,' in *The Western University on Trial*, op. cit., p.37.

[36] University courses in the English-speaking world which deal with Goethe's way of science have come and gone in relation to the specialist knowledge of particular scholars. What is true in a general sense is that an understanding of Goethe's scientific contribution has risen markedly alongside a resurgent appreciation of the contributions of philosophers like Schelling and Novalis who both were connected with the work of Goethe.

[37] C.H. Haskins, *The Rise of the Universities*, Transaction Publishers, New Brunswick, 2002, pp.41-42.

[38] Ibid., p.70.

[39] The meaning of the subjects within the *quadrivium*, as conceived by Plato, was quite different from how we understand it today. Arithmetic was the study of the order of number; geometry the order of space as number in space; music (harmonic theory) as number in time; astronomy as number in space and time. See J. Martineau (ed.), *Quadrivium*, Wooden Books, Glastonbury, 2010, p.4.

[40] B. Readings, *The University in Ruins*, op. cit., p.56.

[41] Bill Readings develops the idea that the University of Berlin was the model of the modern university. See also H. Richardson (ed.), *Friedrich Schleiermacher and the Founding of the University of Berlin*, Edwin Mellen Press, Lewiston, 1991.

[42] See B. Readings, *The University in Ruins*, op. cit., pp.30-32.

[43] A. Gare, 'Democracy and Education: Defending the Humboldtian University and the Democratic Nation-State as Institutions of the Radical Enlightenment', in *Concrescence*, Vol. 6, 2005, p.19.

[44] F.W.J. Schelling, *On University Studies*, (trans. E.S. Morgan), Ohio University Press, Ohio, 1963.

[45] Ibid., p.36.

[46] Ibid., p.36.

[47] See R.J. Richards, *The Romantic Conception of Life*, The University of Chicago Press, Chicago, 2002, p.162.

[48] Ibid., pp.133-134.

[49] Ibid., p.164.

[50] See H. Richardson (ed.) *Friedrich Schleiermacher and the Founding of the University of Berlin*, op. cit., p.35.

[51] R. Steiner, *Nature's Open Secret*, op. cit., p.22.

[52] Ibid., p.25.

[53] C. Wellmon, *Becoming Human*, The Pennsylvania State University Press, University Park, 2010, p.237.

[54] Ibid,. p.249.

[55] Ibid., p.138.

[56] Ibid., p.269.

[57] Ibid., p.212.

[58] W. von Humboldt, 'On the Historian's Task' in *History and Theory*, Vol. 6 No. 1, (1967), pp.57-61 (also at www2.southeastern.edu/Academics/Faculty/jbell/humboldt.pdf).

[59] Ibid., p.61.

[60] Ibid., p.64.

[61] Ibid., pp.58-9.

[62] Ibid., p.71.

[63] See for example Arran Gare, 'Democracy and Education: Defending the Humboldtian University and the Nation-State as Institutions of the Radical Enlightenment,' *op. cit.*, pp. 3-27, and Bill Readings, *The University in Ruins*, op. cit.

[64] For a key example, see the University of Adelaide's (Australia) new *Beacon of Enlightenment* strategic plan, which introduces key elements of the Humboldtian methodology—specifically, research possibilities for undergraduate students, and small group learning experiences rather than lectures. See www.adelaide.edu.au/VCO/beacon/beacon-of-enlightenment.pdf.

[65] Frederick Beiser writes on Wilhelm von Humboldt's views on the role of the state and the university: 'The end of human beings is not happiness, however, still less the accumulation of property. Rather, it is the realization of their characteristic powers, the development of all intellectual, moral, and physical powers into a harmonious whole . . . The state must be a *Bildungsanstalt*, and institution for the development of humanity'. F. Beiser, *Enlightenment, Revolution and Romanticism: The Genesis of Modern German Political Thought*, 1790-1800, Harvard University Press, Cambridge, 1992, p.131.

[66] W. von Humboldt, *Humanist Without Portfolio*, Wayne State University Press, Detroit, 1963, pp.125-132.

[67] Ibid., p.134.

[68] See D. Fallon, *The German University: A Heroic Ideal in Conflict with the Modern World*, Colorado Associated University Press, Boulder, 1980, p.51.

[69] B. Readings, *The University in Ruins*, op. cit., p.54.

[70] Ibid., p.32.

[71] Ibid., p.169.

[72] From J.W. von Goethe, *Roman Elergies I*, trans. by A. S. Kline, https://www.poetryintranslation.com/PITBR/German/Goethepoems.php#anchor_Toc74652110

[73] C.H. Haskins, *The Rise of the Universities*, op. cit., p.xxxi.

[74] Ibid., p.xxviii.

[75] In the key passage from the *Republic* where the three parts of the human soul—the thinking, feeling and willing—are 'tuned' like the notes of a scale. This was the ideal for the perfect human being. See Plato, *The Republic*, Penguin, Harmondsworth, 1974, p.221.

[76] R. Steiner, *A Modern Art of Education*, op. cit., p.45.

[77] Ibid., p.53.

[78] See M.T. Cicero, *On the Ideal Orator* (trans. by J. May & J. Wisse), Oxford University Press, Oxford, 2001.

[79] R. Steiner, *A Modern Art of Education*, op. cit., p.43.

[80] It is this ideal which works within his school educational impulse, often called 'Waldorf education', the first school of which came into existence in his own time.

[81] F.W.J. Schelling, *On University Studies*, op. cit., p.7.

[82] R. Steiner, *Youth and the Etheric Heart*, SteinerBooks, Great Barrington, 2007, p.15.

[83] R. Steiner, *Awake! For the Sake of the Future*, SteinerBooks, Great Barrington, 2015, p.19.

[84] See C.H. Haskins, *The Rise of Universities*, op. cit., p.38-53.

182 THE UNIVERSITY AT THE THRESHOLD

85 One of the first great works on higher education in the Western tradition was Plato's *Republic*. In this he explains how the ideal education leads to the formation of a human being whose nature is temperate and 'perfectly adjusted' and whose actions are good. See Plato, *The Republic*, op. cit., p.221. Wilhelm von Humboldt, the founder of the first modern university in Berlin, had something similar to say many centuries later: 'The true end of Man, or that which is prescribed by the eternal and immutable dictates of reason, and not suggested by vague and transient desires, is the highest and most harmonious development of his powers to a complete and consistent whole'. (W. von Humboldt, *The Limits of State Action*, Liberal Fund Inc., Indanapolis, 1969, p.10.)

86 J. Naydler, *Goethe on Science*, op. cit., p.102.

87 R. Steiner, *West and East: Contrasting Worlds*, op. cit., p.176.

88 Plato does this in both the *Phaedrus* and *Philebus* dialogues.

89 See for example P. Palmer & A. Zajonc, *The Heart of Higher Education*, Jossey-Bass, San Fransisco, 2010, which outlines forms of meditative 'contemplative enquiry' in relation to colleges and universities, largely inspired by the Tibetan Buddhist tradition.

90 J. W. von Goethe, *The Collected Works, Vol. 12, Scientific Studies*, ed. and trans. D. Miller, Princeton UP, New Jersey, 1988, p.39.

91 A. Schopenhauer, *The World as Will and Representation*, vol. I, quoted in C. Janaway (ed.), *The Cambridge Companion to Schopenhauer*, Cambridge University Press, Cambridge, 1999, p.214.

92 See R. Steiner, *The Sun Mystery in the Course of Human History*, Rudolf Steiner Publishing Co, London, 1955, pp.7-8.

93 There are other ways of conceiving of the outer and inner constitution of the human being; for example, Rudolf Steiner speaks of the physical body, the etheric body (body of formative forces), the astral body (sentient body) and the spirit (or 'I').

94 J.W. von Goethe, *Italian Journey*, op. cit., p.153.

95 See for example the series developed by the Bolk Institute in Holland that 'demonstrate how the insights of current biomedical science can be broadened using the Goethean phenomenological method'. Titles include: Anatomy, Physiology, Immunology, Pharmacology, The Healing Process, Respiratory System Disorders and Therapy, Depressive Disorders and Wholeness in Science: A methodology for Pattern Recognition and Clinical Intuition. G. van der Bie et al, *Bolk's Companions on the Fundaments of Medicine*, Louis Bolk Institute, Hoofdstraat, 2012.

96 The application in the humanities of Goethean scientific methods began with the research of Wilhelm von Humboldt in the areas of history, anthropology and linguistics; see Chapter 3. An annotated bibliography, relevant up to 1987, is provided in F. Amrine et al (eds.), *Goethe and the Sciences: A Reappraisal*, D. Reidel Publishing Company, Dordrecht, 1987.

97 In the area of physics, and taking a practical/pedagogical Goetheanistic approach, the following could be referred to: G. Maier, *The Optics of Visual Experience*, Adonis Press, Hillsdale, 1986 and J. Kuhl, *Rainbows, Halos, Dawn and Dusk*, Adonis Press, Hillsdale, 2015.

98 See R. Steiner, *Nature's Open Secret*, op. cit., pp.154-158.

99 Ibid., p.153.

100 See for example the journal *Mensch + Architektur* (Man and Architecture): www.mensch-und-architektur.org

101 See http://wn.rsarchive.org/Articles/GA031/English/UniEdu_index.html

102 R. Steiner, *The Spirit of Fichte Present in our Midst*, Rudolf Steiner Publishing Co., London, p.29.

103 J. Naydler, *Goethe on Science*, op. cit., p.78.

104 C. Root, 'Conversation Between Friends: An Inspiration for Goethe's Phenomenological Method', *In Context*, No. 15 Spring, 2006, pp. 19-22, (see publications of the Nature Institute at: www.natureinstitute.org).

105 Quoted in A. Wiedmann, *Romantic Art Theories*, Gresham Book, Henley-on-Thames, 1986, p.84.

106 P. Goodman, *Compulsory Mis-education and The Community of Scholars*, Vintage, New York, 1964.

107 J.W. von Goethe, *Scientific Studies*, op. cit., p.156.

108 J. Naydler, *Goethe on Science*, op. cit., p.58.

109 R. McKeon, *The Basic Works of Aristotle*, Random House, NY, 1941, p.554.

110 J. W. von Goethe, *Scientific Studies*, op. cit., p.156.

111 J. Naydler, *Goethe on Science*, op. cit., p.57. Kant's views on attraction and repulsion as the two fundamental forces of matter were first expounded in his *Physical Monadology* and later in *Metaphysical Foundations of Natural Science*; here he asserts that both forces together determine the existence of matter, that if there were only attraction matter would coalesce to a point and space would be empty and if there were only repulsion matter would disperse itself to infinity.

112 R. Richards, *The Romantic Conception of Life*, op. cit., p.139.

113 Ibid., p.139.

114 See J. Adler, 'The Aesthetics of Magnetism: Science, Philosophy and Poetry in the Dialogue between Goethe and Schelling', in *The Third Culture: Literature and Science*, Walter de Gruyter, Berlin, 1997, pp.75-6.

115 From F. Schelling, *Further Presentations from the System of Philosophy* (1802), found in M.Vater & D. Wood (eds.), *The Philosophical Rupture between Fichte and Schelling: Selected Texts and Correspondence (1800-1802)*, SUNY Press, NY, 2012, p.169.

116 B. Matthews, *Schelling's Organic Form of Philosophy*, State University of New York Press, NY, 2011, p.166.

117 In his *Freedom Essay* (1809) Schelling shows that the expression 'God is all things' does not mean that there is no difference between God and all things (which is

pantheism). In this sentence 'is' is the copula, mediating both identity and difference. See M. Thomas, 'The Mediation of the Copula as a Fundamental Structure in Schelling's Philosophy,' *Schelling-Studien*, 2 (2014), pp.20-39.

[118] See D. Nassar, 'From a Philosophy of Self to a Philosophy of Nature: Goethe and the Development of Schelling's *Naturphilosophie*' in *Archiv für die Geschichte der Philosophie* 92 (2010).

[119] Ibid.

[120] R. Steiner, *Goethe's World View*, Mercury Press, Spring Valley, 1985, pp.58-59.

[121] The most heightened form of Earth is the ovule, the seed-potency raised on the growing stem and now resting in the darkened 'Earth-like' protection of the ovary. The most heightened form of Sun is the pollen, ripened on the stamen and—either wind-blown or insect-carried—rising beyond the plant into sun-filled heights before descending to unite with the ovule.

[122] See the discussion about this point in W. Pelikan, *Healing Plants*, Vol. 1, Mercury Press, Spring Valley, 1988, pp.12-13.

[123] Ibid., p.34.

[124] Ibid., pp.34-38.

[125] Two works which do enter into this are H. Bortoft, *The Wholeness of Nature*, op. cit., and H. Proskauer, *The Rediscovery of Colour*, Anthroposophic Press, Spring Valley, 1986.

[126] G.W.F. Hegel, *The Science of Logic*, § 168.

[127] See for example H. Bortoft, *The Wholeness of Nature,* op. cit., pp.191-246.

[128] It is important to note that for colour physics (optics) magenta does not even have a wavelength (it is not in the spectrum of light) and it is a matter of contention how it actually appears to us. Speaking strictly from the theoretical point of view, magenta is not a visible colour at all. Nevertheless, it is considered one of the four primary colours of the subtractive model, used in colour printing. A common explanation of this is that the brain 'comes up with' magenta through combining the sensations of red and blue-violet light. For a documentary which delves into the question of magenta see www.youtube.com/watch?v=MmhSX-TMTtJM. Here is explained that the colour spectrum is only *one phase* of the dynamic appearance of the circle of colours (as revealed by Goethe) and for this reason magenta is said to not exist as a spectral colour.

[129] See E. Kant, *Prolegomena to Any Future Metaphysics That Will Be Able to Present Itself as a Science*, § 13, Remark II.

[130] See M. Warnock, *Imagination*, University of California Press, Berkeley, 1976, pp.13-34.

[131] Ibid., pp.46-47.

[132] See E. Kant, *Critique of Judgment*, § 13, Point 6.

[133] Kant, *Critique of Judgment*, § 3, Kant called this creative thinking the *intellectus archetypus*—he conceived it yet denied it to be a human possibility. This is pre-

cisely the kind of creative thinking which Goethe claimed he practiced in his natural scientific studies.

[134] M. Warnock, *Imagination*, op. cit., p.66.

[135] From F. Schelling, *Concerning the Relation of the Plastic Arts and Nature* (quoted in M. Warnock, op. cit. p.69).

[136] See R. Richards, *The Romantic Conception of Life*, op. cit., p.164.

[137] From William Wordsworth's poem 'Tintern Abbey'.

[138] Quoted in M.L. Jackson, 'The State and Nature of Unity and Freedom' in *Biology and the Foundation of Ethics*, (ed. Jane Maienschein et al), Cambridge Press, Cambridge, 1999, p.107.

[139] Ibid., p.105.

[140] See for example, M.L. Jackson, op. cit. Also, T. Lenoir, 'The eternal laws of form: morphotypes and the conditions of existence in Goethe's biological thought', in F. Amrine, F. J. Zucker, and H. Wheeler (eds.), *Goethe and the Sciences: A Re-appraisal*, op. cit., pp.17-28.

[141] Darwin writes in his *Origin of Species*, Chapter V: 'Thus, as I believe, natural selection will tend in the long run to reduce any part of the organization, as soon as it becomes, through changed habits, superfluous, without by any means causing some other part to be largely developed in a corresponding degree. And conversely, that natural selection may perfectly well succeed in largely developing an organ without requiring as a necessary compensation the reduction of some adjoining part.' He notes, however, the validity of the law of compensation of parts in domesticated species. As regards natural evolution, it couldn't be otherwise, for the idea of natural selection is that it works on specific characters or features of organisms. Darwin's theory was not derived from an understanding of the coordinated unity or wholeness of the organism.

[142] J. Naydler, *Goethe on Science*, op. cit., p.117.

[143] Ibid., p.75.

[144] Ibid., p.71.

[145] Ibid., p.72.

[146] Science education in the form of schooling initiated by Rudolf Steiner, works in another way. It is fundamentally phenomenological, starting with the experience of the phenomenon and through this approaching the theory. The approach in these schools is based on Goethe's way of science.

[147] J. Naydler, *Goethe on Science*, op. cit., p.91.

[148] Goethe said that 'from the mathematician we must learn the meticulous care required to connect things in unbroken succession . . '. (J. Naydler, *Goethe on Science*, op cit., p.81).

[149] The Goethean scientist Andreas Suchantke has drawn attention to the work on colour in floral forms by an earlier Goethean scientist—Wilhelm Troll. Troll found that plants which have flowers in the blue-purple range of colours (inward gesture) tend to be tubular (inward or enclosing) whereas plants which have

flowers in the yellow-red range (outward or radiating) tend to have outstretched petals. Troll found that, among the plants whose flowers have outstretched petals, 80.7% have yellow or white blossoms while 16.9% of them have purple or blue petals. Among the plants that have tubular floral forms, 30.3% have yellow or white blossom while 60.3% of them have purple or blue petals. Suchantke goes on to make sense of correlation of colour gesture and floral form in a way which resonates with Schelling's view that nature's creativity is raised to a higher level of potency (consciousness) in artistic creation; what for the plant remains outward, unconsciously expressed in the physical form of the flowers, is for the human being inward, of a soul nature. See A. Suchantke, *Metamorphosis*, Adonis Press, Hillsdale, 2009, p.54.

150 J. von Goethe, *Italian Journey*, op. cit., p.33.

151 A. Suchantke, *Metamorphosis*, op. cit., p.40

152 Gestures of contraction or expansion, or some in-between condition, relate to the environment in which the plant is growing and to the polarity between Sun and Earth discussed in Chapter 5, between the formative influences of light, air, water and earth. A. Suchantke, *Metamorphosis*, op. cit., p. 44.

153 W. von Humboldt, 'On the Historian's Task', op. cit,, p.71.

154 Obviously historical phenomena are not objects like plants or works of art which can be experienced through the senses; nevertheless the facts of history have a quasi-objective character.

155 W. von Humboldt, 'On the Historian's Task', op. cit., p.67.

156 Ibid., p.70.

157 See R. Steiner, *From Symptom to Reality in Modern History*, Rudolf Steiner Press, London, 1976.

158 Ibid., p.66.

159 Rudolf Steiner developed an historical symptomatology which reflected the phenomenological approaches of both Goethe and Wilhelm von Humboldt. See note 157.

160 For an interpretation and application of Wilhelm von Humboldt's indications, see Rudolf Steiner, *The New Spirituality and the Christ Experience of the Twentieth Century*, Anthroposophic Press, Hudson, 1988, pp.7-9.

161 See R. Steiner, *From Symptom to Reality in Modern History*, op. cit., pp.48-50.

162 J.W. von Goethe, *Scientific Studies*, op. cit., pp.120-121.

163 See G.A. Wells, *Goethe and the Development of Science 1750-1900*, Sijthoff & Noordhoff, Alphen, 1978, pp.19-20.

164 H. Bortoft, *The Wholeness of Nature*, op. cit., pp.104-105.

165 W. Schad, *Understanding Mammals: Threefoldness and Diversity*, Vols. 1 & 2, Adonis Press, Hillsdale, 2018, pp.293-363. See also discussed by Mark Reigner 'Horns, Hooves, Spots and Stripes' in *Goethe's Way of Science*, op. cit., p.195.

166 H. Ritvo, *The Platypus and the Mermaid and other Figments of the Classifying Imagination*, Harvard University Press, Cambridge, 1997, p. 4. See also htt.://sciencecases.lib.buffalo.edu/cs/files/antipodal_mystery.pdf.

[167] T. S. Kemp, *The Origin and Evolution of Mammals*, Oxford University Press, Oxford, 2007, pp.173-4.

[168] W. C. Warren et al, 'Genome analysis of the platypus reveals unique signatures of evolution' in *Nature*, 453, (8 May 2008), pp.175-183. For an example of how the news of the platypus as 'missing link' was broadcast in the general press, see K. Bradford 'Secrets of the Platypus Revealed' in *Sydney Morning Herald*, May 8, 2008. P. Myers, 'Interpreting shared characteristics: The platypus genome', in *Nature Education* 1(1):46, 2008.

[169] 'And it is in fresh water that we find seven genera of Ganoid fishes, remnants of a once preponderant order: and in fresh water we find some of the most anomalous forms now known in the world, as the *Ornithorhynchus* and *Lepidosiren*, which, like fossils, connect to a certain extent orders now widely separated in the natural scale. These anomalous forms may almost be called living fossils; they have endured to the present day, from having inhabited a confined area, and from having thus been exposed to less severe competition'. C. Darwin, *The Works of Charles Darwin Vol. 16, Origin of Species*, New York University Press, NY, 1988, p.87. It actually wasn't until 1884 that it was finally shown that this animal lays eggs.

[170] See the quotation from Friedrich Schelling in Chapter 1.

[171] See 'Energetics and foraging behaviour of the Platypus *Ornithorhynchus anatinus*' by Philip Bethge PhD thesis, www.eprints.utas.edu.au/2326/3/Bethge_PhD.pdf.

[172] J. Holland, '40 Winks?' *National Geographic*, 200 (1), July 2011.

[173] See www.youtube.com/watch?v=dsd7ZfdZcNU.

[174] R. Khamsi, 'Duck-billed platypus boasts ten sex chromosomes' in *Nature*, 25 October, 2004.

[175] C. Holdrege, 'How Does a Mole View the World?' in *In Context* No. 9, (Spring, 2003), pp.16-18.

[176] This was first discussed by Gauthier several decades ago. See J. Gauthier, 'Saurischian monophyly and the origin of birds' in *The Origin of Birds and the Evolution of Flight* (ed. K Padian), Academy of Science, San Francisco, 1986, pp.1-55.

[177] See M. Ebach, J. Morrone, D. Williams, 'Getting Rid of Origins,' in *Theoretical Biology Forum* 99, 2006, pp.357-380. Also M. Ebach, 'Anschauung and the Archetype: The Role of Goethe's Delicate Empiricism in Comparative Biology', *Janus Head*, 8(1), 2005, pp.254-270.

[178] See for example Michael Vater writing the introduction to F. Schelling, *Bruno, Or On the Natural and Divine Principle of Things*, State University of New York Press, NY, 1984, p.14.

[179] Goethe, commenting on the passage where Kant calls the *intellectus archetypus* a human impossibility, writes: 'What should it not ... hold true in the intellectual area that through an intuitive perception of eternally creative nature we may become worthy of participating spiritually in its creative processes? Impelled

from the start by an inner need, I had striven unconsciously and incessantly toward primal image and prototype...' J.W. von Goethe, *Scientific Studies*, op. cit., p.31.

[180] J.W. von Goethe, *Scientific Studies*, op. cit., pp.31-32.

[181] H. Bortoft, *The Wholeness of Nature*, op. cit., p.22.

[182] In particular in J.G. Herder, *Outline of a Philosophical History of Humanity* (1774).

[183] Ibid. Herder spoke of *Volksgeist* (spirit of a people) which had a characteristic *Schwerpunkt* (focal point or 'centre of gravity'). The *Schwerpunkt* of a people is the central social practice from which they cannot be moved because everything depends on it. Herder indicated that this 'centre of gravity' is not understood by the abstract intellect but by 'feeling into' (*Einfühlung*).

[184] Quoted in Ernst Lehrs, *Man or Matter*, Rudolf Steiner Press, London, 1985, p.123.

[185] Quoted in R. Steiner, *Goethe's World View*, op. cit., p.86.

[186] R. Steiner, *The Science of Knowing*, op. cit., p.82.

[187] R. Steiner, *Nature's Open Secret*, op. cit., p.159. Steiner goes on to explain that Goethe wanted to trace back the phenomena of the atmosphere to their causes which lay in the being of the Earth itself. He considered barometric pressure as the archetypal phenomenon and sought to connect everything else to it.

[188] 'Every musical work, every finished tonal pattern, grows out of a seed that lies hidden and yet reveals itself in the pattern, which constantly leaves it behind and at the same time carries it forward. Although the "seed"—the primal form— is entirely dissolved in the pattern it is the fundamental law governing the organization of the pattern. One and the same law determines the form and places of each of its parts and the way they are knit together into a whole. The process by which the finished pattern is gradually produced from the seed is called "transformation".' V. Zuckerkandl, *Man the Musician*, Princeton University Press, NY, 1973, p.171. See the influential way of analysis by Heinrich Schenker which discerns three layers of musical formation—the *Ursatz* working in the background, which is elaborated to some degree in the middleground, and to the greatest degree in the foreground, forming the 'surface' of the music.

[189] See G. van der Bie, *Wholeness in Science*, Louis Bolk Institute, Hoofdstraat, 2012, pp. 83-4. This has enormous significance in relation to our understanding of evolution. The Darwinian view is that each part of an organism evolves independently—for example, the beak of a bird evolved independently from its feet.

[190] This expression comes from the epilogue to *Beaumont and Fletcher's Honest Man's Fortune*. John Fletcher (1579—1625) was a Jacobean playwright.

[191] Rudolf Steiner outlines Goethe's whole exploration into animal morphology, beginning in his mid-twenties with his studies of *Essays on Physiognomy* by Swiss poet and philosopher Johann Lavater and the writings of Aristotle on physiognomy (R. Steiner, *Nature's Open Secret*, op. cit., pp.22-41).

[192] R. Steiner, *Nature's Open Secret*, op. cit., p.27.

[193] J. Verhurlst, *Development Dynamics in Humans and Other Primates: Discovering Evolutionary Principles through Comparative Morphology*, Adonis Press, Hillsdale, 2003.

[194] S. Kneller, *Inventory of the Universe*, Wheatmark, Tucson, 2016.

[195] P. Rosch (ed.), *Bioelectricmagnetic and Subtle Energy Medicine*, CRC Press, London, 2015, p.461.

[196] E.E. Ruppert et al, 'Annelida', in *Invertebrate Zoology*, (7th ed.), Brooks/Cole, pp.414-420.

[197] See www.en.wikipedia.org/wiki/Wedge-tailed_eagle

[198] J. Serpell (ed.), *The Domestic Dog*, Cambridge University Press, Cambridge, 1995, p.107.

[199] J. Napier writes: 'The human hand displays an exceptional degree of primitiveness—an astonishing conclusion if we consider that it is capable of specialized movements and exceptional sensitivity, precision, subtlety, and expressiveness.' Quoted in J. Verhulst, *Developmental Dynamics*, op. cit., p.124.

[200] Ibid., p.124.

[201] Steiner first wrote about his understanding of the threefold nature of the human organism in his *Riddles of the Soul* (1917).

[202] W. Schad, *Understanding Mammals: Threefoldness and Diversity*, op. cit.

[203] Ibid., Vol. 1.

[204] Discussed, for example, in R. Dawkins, *The Blind Watchmaker*, W.W. Norton, 1986.

[205] www.en.wikipedia.org/wiki/Rodent

[206] R.W. Braithwaite, 'Brown Antechinus', in R. Strahan (ed.), *The Mammals of Australia*, Reed Books, pp.94-97.

[207] W. Schad, *Understanding Mammals*, op. cit., p.1024.

[208] Ibid., p.295.

[209] Ibid., p.246. The European long-tailed mouse has these features also.

[210] Ibid., p.244.

[211] R.W. Braithwaite, op. cit.

[212] W. Schad, *Understanding Mammals*, op. cit., p.1135.

[213] Ibid., p.998.

[214] Ibid., p.1002.

[215] L. Werdelin, 'Comparison of Skull Shape in Marsupial and Placental Carnivores,' in *Australian Journal of Zoology*, 34 (2), 1986, pp.109-117.

[216] R. Dawkins, *The Blind Watchmaker*, op. cit., p.105.

[217] J. Dixon, www.environment.gov.au/system/files/pages/a117ced5-9a94-4586-afdb-1f333618e1e3/files/20-ind.pdf

[218] Ibid.

[219] See M. Riegner, 'Horns, Hooves, Spots and Stripes,' op. cit., p.187.

[220] W. Schad, *Understanding Mammals*, op. cit., p.675.

[221] Ibid., p.675.

[222] Ibid., p.675.

[223] J. Dixon, op. cit. p.9.

[224] Ibid., p.13.

[225] W. Schad, *Understanding Mammals*, op. cit., p.297.

[226] Dixon, op. cit., p.14.

[227] Ibid.

[228] Kangaroo molars move forward as they are ground down, and fall out, replaced by teeth that grow in the back. This is known as polyphyodonty and otherwise occurs only in elephants and manatees. www.en.wikipedia.org/wiki/Kangaroo

[229] R. Steiner, *World Economy*, Anthroposophic Press, NY, pp.201-5.

[230] Victor Hugo, *Les Misérables*, Part V, Book 1 (1862).

[231] C. Houghton Budd, *Prelude in Economics*, New Economy Publications, 1999, p.11.

[232] R. Steiner, *The Renewal of the Social Organism*, Anthroposophic Press, Spring Valley, 1985, p.126.

[233] See G. R. MacLay, *The Social Organism: A Short History of the Idea that Human Society May be Regarded as a Gigantic Living Creature*, North River Press, Croton-on-Hudson, 1990.

[234] This view was upheld by influential theorists of the time, such as Christian Wolff.

[235] W. von Humboldt, *The Limits of State Action*, op. cit.

[236] See P. Franco, *Hegel's Philosophy of Freedom*, Yale University Press, New Haven, 1999, Chapters 5 & 6.

[237] Spencer thought that this division of races and classes arose through natural selection and did so after reading Darwin's *Origin of Species*. It was Spencer, not Darwin, who coined the term 'survival of the fittest', first used in his *Principles of Biology* (1864).

[238] S. Kesebir, 'The Superorganism Account of Human Sociality: How and When Human Groups are like Beehives', *Personality and Social Psychology Review*, published online, December 2011.

[239] J. Habermas, *Legitimation Crisis*, (trans. T. McCarthy), Beacon Press, Boston, 1975, pp.5-6.

[240] Steiner's ideas on the threefold social organization are spread through many lectures he gave in the last decade of his life. His one written work on the subject is *Toward Social Renewal: Basic Issues of the Social Question* (first entitled *The Threefold Social Order*).

[241] R. Steiner, *Freedom of Thought and Societal Forces*, op. cit., pp.98-99.

[242] Today the idea of 'natural capital' has come into existence, implying that trees, minerals etc, have some kind of economic value. However, such is what Rudolf Steiner has called 'apparent' as opposed to 'real values'. (R. Steiner, *World Economy*, op. cit., p.97). Value only arises when nature is worked on in some way, through human labour. When land is owned (capitalized) the value it appears to

obtain in fact amounts to a storing and freezing of capital, unlike in a bank where capital can be constantly put to use.

[243] G. Lamb & S. Hearn (eds.), *Steinerian Economics: A Compendium*, Adonis Press, Hillsdale, 2014, p.194.

[244] The Organization for Economic Co-operation and Development defines it as 'the ratio of a volume measure of output to a volume measure of input.', OECD Manual: *Measuring Productivity: Measurement of Aggregate and Industry-Level Productivity Growth*, 2002.

[245] C. Houghton Budd, *Prelude in Economics*, op. cit., p.22.

[246] See for example G. S. Becker, *Human Capital: A Theoretical and Empirical Analysis, with Special Reference to Education*, University of Chicago Press, Chicago, 1964, 1993.

[247] J. Weatherford, *The History of Money*, Three Rivers Press, NY, 1997.

[248] C. Houghton Budd, *Prelude in Economics*, op. cit., p.49.

[249] G. Lamb & S. Hearn (eds.), *Steinerian Economics*, op. cit., p.227.

[250] Production out of an inner impulse, the activity we call 'creative expression', arising from the need to bring something to outer expression which arises from inner experience, belongs to the cultural-spiritual sphere of society. Michael Spence makes clear the distinction between economic and cultural production in his *After Capitalism*, Adonis Press, Hillsdale, 2014, p.87.

[251] Houghton Budd writes: 'Ultimately, even the dispensing of capital is an untruth. It can only be drawn upon . . . If we wanted to reach the source of true social action within economics, we would make capital available in accordance with our individual assessments of our own value'. op. cit. p. 28. This, clearly, is an ideal. In that direction we may conceive what Rudolf Steiner suggests is a feasible and necessary step: that the cultural-spiritual sphere primarily (but in conjunction with the political-rights sphere) administer the dispensation of capital. Capital, Steiner indicates, when understood as a community asset rather than private property, should only be placed in the hands of an entrepreneur as long as they can be productive with it for the benefit of the community—beyond which point it is passed on to other capable and productive people without charge. See R. Steiner, *The Social Future*, Anthroposophic Press, NY, 1972, pp.118-119 and R. Steiner, *Towards Social Renewal*, Rudolf Steiner Press, Forest Row, 1999, pp.79-80.

[252] R. Steiner, *The Renewal of the Social Organism*, op. cit., p.101.

[253] C. Houghton Budd, *Prelude in Economics*, op. cit., p.27.

Picture Credits

A NOTE FROM RUDOLF STEINER PRESS

We are an independent publisher and registered charity (non-profit organisation) dedicated to making available the work of Rudolf Steiner in English translation. We care a great deal about the content of our books and have hundreds of titles available – as printed books, ebooks and in audio formats.

As a publisher devoted to anthroposophy...

- We continually commission translations of previously unpublished works by Rudolf Steiner and invest in re-translating, editing and improving our editions.

- We are committed to making anthroposophy available to all by publishing introductory books as well as contemporary research.

- Our new print editions and ebooks are carefully checked and proofread for accuracy, and converted into all formats for all platforms.

- Our translations are officially authorised by Rudolf Steiner's estate in Dornach, Switzerland, to whom we pay royalties on sales, thus assisting their critical work.

So, look out for Rudolf Steiner Press as a mark of quality and support us today by buying our books, or contact us should you wish to sponsor specific titles or to support the charity with a gift or legacy.

office@rudolfsteinerpress.com
Join our e-mailing list at www.rudolfsteinerpress.com

RUDOLF STEINER PRESS